MULTIPLE-CHOICE AND FREE-RESPONSE QUESTIONS IN PREPARATION FOR THE AP BIOLOGY EXAMINATION

(FIFTH EDITION)

By

Glenn Hartman
Archmere Academy
Claymont, Delaware

and

Jennifer Pfannerstill
Tomahawk High School
Tomahawk, Wisconsin

 **D&S MARKETING SYSTEMS, INC.**
1205 38th Street • Brooklyn, NY 11218

w w w . d s m a r k e t i n g . c o m

ISBN # 0-9787199-6-4

Preface

The College Board Advanced Placement Biology Examination consists of an 80-minute multiple-choice section composed of 100 items and a 90-minute free-response section that requires students to respond to four mandatory questions. Students are allotted ten minutes to plan answers to the free response questions before they begin writing.

This book is intended to provide students and teachers with a comprehensive set of review materials to help students prepare for the multiple-choice portion of the exam. This edition consists of review material, organized themes, with twenty multiple-choice questions and two free-response questions, organized with each of the eight themes. There are also three full length multiple-choice examinations, making a total of 460 challenging practice questions and twenty free-response questions. Review material for the twelve suggested labs is included and there are lab based questions throughout the book. This book also contains item analysis pages for each of the multiple-choice practice examination to assist the student in determining areas of strength and areas needing additional review.

The Teacher's Manual contains complete explanations for each sample question which can be shared with students at the teacher's discretion, as well as rubrics for the 20 free-response questions in the book.

The writing of this 5th edition of the AP biology review book has been a a very rewarding process. We would both like to recognize our families for their support and generous giving of time to allow for the creation of this document, our current and past students who patiently tried out many questions, and other AP readers, including Bernie Shellem, Brent Thomas, Elizabeth Cowles and Paula Phillips who provided support and helpful suggestions and comments. Special appreciation is in order to Mr. Franklin Bell for his superb editing and suggestions for making the writing process, and the final work, of the highest quality and to the textbook publishers who granted permission for the use of their illustrations.

All communications concerning this book should be addressed to:

D&S Marketing Systems, Inc.
1205 38th Street
Brooklyn, NY 11218
www.dsmarketing.com

TABLE OF CONTENTS

AP Biology

Survival Guide for Students

What is AP Biology?

The course in AP Biology is designed to be a college level survey in biological sciences, touching on the mainstays of biological science by using a thematic approach. These unifying themes include:

I.   Science as a Process
II. Evolution
III. Energy Transfer
IV. Continuity and Change
V.   Relationship of Structure to Function
VI. Regulation
VII. Interdependence in Nature, and
VIII. Science, Technology, and Society

These themes, while always the driving force behind the conceptual framework of AP Biology, are, as of late, becoming the new focus for the AP Biology Development Committee in their quest to improve the AP Biology course and the learning of AP Biology students. Making thematic connections will encourage higher-level thinking and establish more meaningful learning for the students.

Throughout this book, we have identified several of the content areas below as part of specific themes; this is not to say that particular topics cannot be addressed in other themes. Rather, it is a starting point with a solid thematic approach for you to identify connections between content areas. It is important to remember that while this is an objective based test, questions are designed to assess conceptual knowledge, not simply discrete facts. If teachers are able to emphasize these concepts and apply them to the themes, students will not only be more prepared for the test, they will become more knowledgeable scientists.

The content areas below should be covered in this introductory level college biology course; the percentages are indicative of both the length of time within the course for each topic as well as the appropriate number of multiple-choice questions on the exam.

Unit and Subtopics	Percentage
Molecules and Cells	**25%**
Chemistry (water, organic molecules, energetics, enzymes)	7%
Cells (cell transport, cell structure)	6%
Photosynthesis	4%
Cellular Respiration	4%
Cell Division (phases and regulation)	4%
Genetics and Evolution	**25%**
Heredity (meiosis and gametogenesis, eukaryotic chromosome, inheritance patterns)	8%
Molecular Genetics (RNA/DNA structure and function, gene regulation and mutation, viral structure and replication, and biotechnology)	9%
Evolution (evidence and mechanisms for evolution and early evolution of life)	8%
Organisms and Populations	**50%**
Diversity of Organisms (evolutionary relationships, diversity of life, and phylogenetics)	8%
Plant Structure and Function (reproduction, growth, development; structural, physiological, and behavioral adaptations, and environmental responses)	32%
Ecology (populations, communities, ecosystems, and global issues)	10%

What is the AP Biology Exam?

The AP Biology exam is taken in early May and consists of two parts. The multiple choice section is administered first and consists of 100 questions, each with five possible choices, making up 60% of the student's final score. Students are given 80 minutes to answer the questions. Students are not limited, as in many standardized tests, from looking at other portions of the multiple-choice section throughout the 80-minute testing period. However, once that section has been completed and time has expired, students will not have access to the multiple-choice section while the free response portion of the exam is administered.

In the 100 multiple-choice questions, there are three primary types of questions. The first 30-40 questions are individual scattered questions that stem from any area of biology. The questions will require specific content knowledge as it relates to one of the primary AP biology themes. The next set of questions is arranged in groups where there will be a drawing, experimental description, or some information that can be interpreted. Answer choices may apply to a group of 3-5 questions. The final section of questions follows written descriptions, graphs, or experiments and requires analysis of that information. These tend to be individual analysis questions rather than related groupings as in the previous section.

Topically, the questions should be divided so that 25% relate to molecules and cells, 25% relate to heredity and evolution, and 50% relate to organisms and populations. These content areas are divided among the three types of exam questions. Also, questions from the assigned AP biology laboratory experiments as they relate to these topics appear throughout the multiple-choice section of the exam.

The free response part of the exam consists of four essay questions spanning ninety minutes and comprising 40% of the final score. As with the multiple choice section, there is a breakdown of topics represented in the free response section: one essay question will assess understanding in *Molecules and Cells*, one question will assess understanding in *Genetics and Evolution*, and two questions will assess understanding *Organisms and Populations*. Objectives from the College Board's twelve required labs will be addressed in at least one of the questions, and may ask students to bridge concepts from more than one lab in the quest for a deeper understanding of biological concepts as they are applied.

Free response questions very frequently have *directional* words in them that students should notice. These words tell the students exactly what type of knowledge, and therefore response, is required. Directional words include: describe, discuss, compare, contrast, explain, list, etc. Free response questions are also usually designed in multiple parts. Unlike other AP tests, all free response questions on the AP Biology exam are graded; there is no choice. However, within each question, there may be individual choice to answer particular parts of a question.

In an effort to promote more organized responses, the College Board recently changed the format to include an additional reading time of ten minutes. The first ten minutes of this test is considered a "reading period" in which students only have access to the questions and scratch paper, but no test answer booklet. This is designed to be a brainstorming time where students are encouraged to read the questions clearly, begin outlining their responses and perhaps decide which question they would like to answer first. After this, the ninety minute free response portion of the exam begins. Questions are reprinted in the answer book and there is substantial space for student responses in each section. Students are also allowed to have their brainstorming work from the reading period with them.

How is the Exam graded?
Exams are given a holistic score of 1, 2, 3, 4, or 5 based on the student's multiple choice score (60%) and free response score (40%). In the multiple-choice section, students earn one point for each correctly answered multiple-choice question. While students are not penalized for leaving a question unanswered, they do lose ¼ of a point for each question answered incorrectly. The free response portion of the exam is graded by nearly 400 readers who design standards by which to grade the exam and objectively and consistently apply the rubric to each response. A maximum of ten points may be awarded for each essay. Points are then collated for each section and converted by the Chief Faculty Consultant to a holistic score of 1, 2, 3, 4, or 5 with 5 being extremely well qualified, 4 being well qualified, 3 being qualified, 2 being possibly qualified, and no recommendation for a score of 1.

Tips for Completing the Multiple-Choice Section:

1. Since there is a small penalty for incorrect answers (a loss of ¼ of a point for each incorrect answer), it is not best to guess haphazardly in this section. With five answer choices to each question, guessing should be avoided if no answer choices can be eliminated. If at least two answer choices can be eliminated, guessing out of the remaining three is recommended. If only one answer choice can clearly be eliminated, it should be the discretion of the test taker if guessing is chosen. There isn't any statistical advantage to guess or avoid guessing when four answer choices remain viable.

2. Bring several number 2 pencils with very clean erasers. Extraneous markings on the scantron test can interfere with the accuracy of the scanning equipment.

3. No calculators are permitted in this section. Any questions that require mathematical analysis should be able to be done readily without the aid of a calculator.

Tips for Completing the Free-Response Section:

1. Read each question thoroughly. Pay close attention to the verbs in the questions: identify if you are supposed to list, describe, or explain something and then focus writing of the essay around that action.

2. Identify the various components to the question. Divide questions that have multiple parts into separate sections even if not specifically asked for in the question because it makes the reading of the essay easier and it easier to see how you addressed each section in your response.

3. Pay attention to words like "and" and "or." It is important to answer only as many sections or give as many examples as directed to provide. Do not provide extra examples when a set number is requested. If no limit on example numbers is provided, feel free to elaborate with additional examples.

4. If a question asks you to explain 3 of 4 terms, only explain three. Any information you give on a fourth term will not be awarded points and will only be consuming valuable time.

5. Use the ten-minute organization time to briefly outline each response in the green question book. This organization is essential to formulate a solid, well thought free-response.

6. Write in blue or black pen. Pencil is hard to read. Colored ink is distracting and not as clear to the reader.

7. Use the best and clearest handwriting that you can. Readers do not want to have to reread a section of a response several times to prevent missing some point-worthy comments. Printing may be recommended since it tends to be the clearest.

8. Points are given for valid information with explanations. So, elaborate on your responses with explanations and examples. Often single related-words or unexplained references will not be allocated any points.

9. Use time wisely. You have 90 minutes to sufficiently answer 4 free-response questions. If you organize your thoughts during the ten-minute organization time, you have 22 minutes of writing time for each free response answer. Please, use that time. Do not provide quick, incomplete responses. These incomplete or unexplained responses are not generally awarded many points.

10. Compose your response in brief essay format. You must write in complete sentences and paragraphs to receive points. Bulleted answers and lists will not be awarded points. However, refrain from writing verbose essays that reword the same comment several times. You do not need to waste time writing formal essay introductions and summaries. All of the detailed, well-explained points will be awarded for information in the body of a traditional essay.

11. Keep the essay responses focused. If a question or part of a question asks you to focus on a specific component of a topic, stay focused on that component. You will not receive points for additional information on the concept that are not requested in the question. For example, if a question asks you to describe how genetic variation occurs in meiosis, do not waste time describing every event of each phase of meiosis. That information is not directly related to the posed question and will not earn points and will again consume valuable time.

12. When answering a lab-related free-response question, provide all material that is requested. For graphs, provide titles and clearly defined and labeled axes. For experimental designs, describe plausible examples with clearly noted controls and experimental groups, standard conditions, replication, and data collection and analysis.

How can teachers and students use this book?

This book is divided into several sections within the teacher manual and student study guide. Due to the ever-changing nature of the AP Biology course and its potential redesign, this book has been formatted to incorporate and follow a thematic approach. Within each of the themes identified by the College Board as unifying themes for an introductory biology course, there is specific content material. These content areas have been sorted to accompany appropriate themes. However, many topics can be addressed in several themes; we have only suggested possible connections. The content is preceded by a ***boldfaced*** conceptual statement designed to demonstrate a relationship within each theme. Within each section, there is a content review section, although not exhaustive, multiple choice and free response questions, and answers to the questions (teacher's manual only). There are also three full-length exams with annotated answers (teacher's manual only) at the conclusion of the book. By omitting the answers in the student manual, teachers may choose to use the questions as an assessment tool. However, teachers may also choose to provide those answers to their students, thus providing context for the content material. Students and teachers alike can benefit from the annotated answers.

Theme One

Science as a Process

As with any science, all content and knowledge on biological principles stems from hundreds of years of on-going scientific research. Every fact and concept now widely accepted was once unknown and discovered through scientific research. So, every concept in biology can be tied to scientific process. In this section, some historical information and the AP Biology labs are used as a foundation for how biological material can be expressed through the process of science.

Exceptional experimental design is crucial in developing advanced understanding of biological principles. Proper identification of independent and dependent variables, using appropriate controls, analyzing data statistically, making quality observations and understanding how error may be addressed within the laboratory all influence our understanding.

- *In order to understand the foundations and principles of evolution, it is important to analyze the work and contributions of several scientists. The collected research of these individuals led to the understanding of the evolutionary process.*

Scientist	Contribution to Evolutionary Thought and Understanding
Carolus Linnaeus	• Founder of Taxonomy • Developed the **Binomial Nomenclature** Naming System
James Hutton	• Profound change can happen to the earth's surface due to slow gradual changes (**gradualism**)
Jean-Baptiste de Lamarck	• Suggested that living organisms change as their environments change • Erroneous conclusion that organisms can pass acquired traits on to offspring
Thomas Malthus	• Organisms have a capacity to overreproduce and experience pressure due to resource limitations
Georges Cuvier	• Fossils varied between rock strata • Proposed catastrophism as the reason for this variance—sudden large disturbances caused great changes in organism diversity
Charles Lyell	• Earth is older than previously thought • **Uniformitarianism**—the same features that alter the earth's surface today affected the earth's surface in the past with a similar impact and at a similar rate

1

Charles Darwin	• Collected evidence about flora and fauna around the globe to assess how the diversity appeared • Increases in population sizes result in a struggle for survival within a population • The unequal reproduction rates will cause the favorable traits to accumulate over time
Alfred Wallace	• Populations experience change due to environmental pressures.

• *The knowledge of genes, chromosomes, and heredity is due to contributions of many notable scientists. Each scientist was able to add a new piece of information to the growing bed of knowledge.*

Scientist	Contribution to the Understanding of Inheritance and Heredity
Gregor Mendel	• Analyzed 7 traits with garden peas • Noticed that each individual has two alleles for each trait • The alleles for one trait assort independently for the alleles for another trait • When two different alleles are present, one appears to be dominant over the other and is always expressed
Walter Sutton	• Helped develop the **Chromosomal Theory of Heredity**; Genes for traits are located on chromosomes
Thomas Hunt Morgan	• Discovered that sometimes alleles for different traits do not assort independently and can be linked because they exist on the same chromosome • Discovered that sex-linked traits appear in different rates in males and in females because males need only one recessive allele to express the recessive trait

• *The work of many scientists led to great discoveries is in the area of understanding DNA to be the genetic material and isolating the structure and role of DNA in cells.*

Scientist	Contribution to Understanding the Function and Structure of DNA
Frederick Griffith	• Transformation of genetic material is possible through use of two pneumonia strains
Oswald Avery, Maclyn McCarty, Colin MacLeod	• Repeated Griffith's experiments to find that the transforming agent was DNA
Alfred Hershey and Martha Chase	• Used bacteriophages to isolate that the DNA from the phage transforms the bacterial host

Erwin Chargaff	• The percentage of the four types of nucleotides is different in different organisms • Percent of adenine nucleotides is the same as the percent of thymine nucleotides • Percent of cytosine nucleotides is the same as percent of guanine nucleotides
Maurice Wilkins	• Operated a lab where X-ray crystallography was conducted to analyze molecules
Rosalind Franklin	• Took the first X-ray picture of DNA
James Watson and Francis Crick	• Proposed the double-helix model of DNA based upon contributions of earlier scientists • Proposed semi-conservative model of replication, but lacked supporting evidence
Matthew Meselson and Franklin Stahl	• Used isotopes of nitrogen to demonstrate that DNA replication could not be conservative or dispersive • Evidence that DNA replication is semi-conservative
George Beadle and Edward Tatum	• Using bread mold were able to discern that each gene appears to be responsible for making one enzyme that is needed for a biological process

Lab #1: Diffusion and Osmosis Lab

- *By using experimentation, we are able to investigate the tendencies of molecules to maintain equilibrium and the structure and function of cell membranes.*
  - o **Diffusion** is the movement of molecules from an area of high concentration to low concentration.
  - o **Osmosis** is the diffusion of water across a selectively permeable membrane from an area of high water concentration (high water potential) to an area of lower water concentration (low water potential).
  - o **Active Transport** uses energy to move molecules from areas of low concentration to areas of high concentration.
  - o A membrane is **semi-permeable** or selectively permeable if it allows some molecules to cross, but not others. These membranes allow very small molecules like water and other small molecules (IKI and glucose in this experiment) to cross, but large molecules like sucrose and other macromolecules are blocked.
  - o IKI tests for the presence of starch.

Lab #2: Enzyme Catalysis Lab

- *Experimentation allows us to determine optimum environmental conditions for enzymes and promote our understanding of how enzymes interact with substrates and other molecules.*
  - o An **enzyme** is a protein that binds to a reactant in a chemical reaction to increase the rate of the reaction.

o The reactant that the enzyme binds to is called the **substrate**. The enzyme and substrate bind due to their three-dimensional shapes at the **active site**.

o The substrate and enzyme relationship is highly specific and each biological reaction has a specific enzyme that catalyzes that reaction.

o Since enzyme-substrate binding is due to a complementary 3-dimensional shape, environmental factors can affect the binding of the reaction.

o An increase in temperature will cause molecules to move faster and will increase the rate of reaction. However, extreme heat will cause the enzyme protein to denature or lose its three-dimensional shape and lose its function.

o Each enzyme is most efficient at a specific pH. Any pH level outside of a range around that optimum will disrupt the bonding that reinforces the enzyme shape and will slow the rate of catalysis.

o **Competitive inhibitors** are molecules that compete with the substrate for the enzyme's active site and interfere with the substrate binding to that active site. This slows the enzyme-catalzyed reaction.

o **Noncompetitive inhibitors** bind to the enzyme outside of the active site and change its three-dimensional shape. This overall shape change causes a shift in the shape of the active site interfering with the substrate binding to that site and slowing the rate of the reaction.

Lab #3: Mitosis and Meiosis Lab

• *Mitosis and meiosis, while both processes involving cell division, result in different outcomes that can be analyzed in the lab.*

o Cells in growth areas of plants (meristems in root and shoot tips) and animals (blastulas and developmental stages) are actively experiencing cell division. In these areas, the amount of cells dividing is much greater than in more mature tissue areas.

o When cells divide, they experience a nuclear division called **karyokinesis** and a cytoplasmic division called **cytokinesis**.

o There are two common types of **karyokinesis**: mitosis that produces somatic cells and meiosis that produces gametes in animals or spores in plants.

o In **mitosis**, the resulting daughter cells are diploid like the parent cell and are genetically identical to the parent cell.

o In **meiosis**, the resulting daughter cells are haploid (half of the chromosomes of the parent cell) and each daughter nuclei is genetically unique.

o The haploid condition of meiotic daughter cells is achieved through one replication during interphase followed by two divisions: meiosis I and meiosis II.

o The unique genetic quality of meiotic daughter cells results due to **independent assortment** (random alignment and separation of paternal and maternal chromosomes) and **crossing-over** (chromatid pieces are exchanged between homologous chromosomes during Prophase I).

Lab #4: Plant Pigments and Photosynthesis Lab

• *Plant tissue contains a variety of pigments that provide color and enhance the light absorption abilities to increase photosynthetic efficiency. By manipulating environmental variables in the lab we are able to analyze the behavior of these pigments in relationship to energy transfer and storage.*
   o Each pigment has different chemical properties and levels of solubility. It is possible to separate these pigments in a separation medium like paper due to the varying solubilities in any solvent.
   o The **R$_f$ value** calculates the distance that pigments migrate in solution and can be used to identify separated pigments.
   o Photosynthetic productivity is most efficient when there is abundant light and the chloroplasts are unboiled.
   o Boiling of chloroplasts denatures the proteins involved in the process of photosynthesis and completely disrupts any photosynthetic activity.
   o A dark environment doesn't allow for the initial light absorption necessary for the light-dependent reactions of photosynthesis.
   o DPIP is a blue-colored chemical that reduces in the presence of high energy electrons and becomes clear. Photophosphorylation will reduce the DPIP and this reduction can be measured and interpreted as photosynthetic efficiency. A reduction in color is indicative of an efficient photosynthetic process.

Lab #5: Cell Respiration Lab

• *All living organisms undergo some type of energy forming process that can be analyzed through experimentation and provide data about energy transfer and storage.*
   o All living organisms require ATP for efficient cell functioning. If no ATP is available, cells must be able to make ATP by converting chemical energy into ATP.
   o Only inactive or dormant cells can be sustained without forming ATP.
   o During cellular respiration, glucose and oxygen are consumed to make ATP. Water and carbon dioxide are the by-products of this process.
   o Cool temperatures can slow the rates of respiration in living organisms.
   o Measuring the amount of oxygen gas consumed is difficult since carbon dioxide gas is produced as a by-product. The reduction of oxygen gas is balanced by the production of carbon dioxide.
   o Potassium hydroxide can be used to absorb the carbon dioxide. This way the overall gas concentration can be used to accurately measure the rate of oxygen gas consumption.

Lab #6: Molecular Biology Lab

- *Genetic recombination produces new phenotypes for offspring through bacterial transformation.*
  - o **Plasmids** are circular pieces of DNA that sit outside of the primary chromosome in bacteria. These extranuclear fragments of DNA can have active genes on them.
  - o Bacteria will accept foreign DNA from the environment through the process of **transformation**. In transformation, protein receptors bind foreign DNA that is then pulled into the cell's interior. This transforming quality provides a level of genetic diversity that is otherwise absent with bacteria.
  - o **Recombinant plasmids** have been engineered to have genes from another source incorporated into the plasmid.
  - o **Restriction enzymes** are used to cut the plasmids and the gene of interest and **DNA Ligase** is used to seal the desired gene into the desired plasmid.
  - o Bacteria can be invoked to undergo transformation by stressing their environments.
  - o Antibiotic resistance genes are often incorporated into the recombinant plasmid to separate transformed bacteria from untransformed bacteria. Transformed bacteria will produce proteins to be resistant to a certain antibiotic and can be grown on a food medium containing that antibiotic. Untransformed bacteria will be killed by the presence of the antibiotic and cannot grow on any media with the antibiotic included.
  - o Restriction enzymes can also be used to cut DNA pieces into small fragments that can be analyzed through separation.
  - o With **electrophoresis**, digested DNA is loaded into a well on a separation medium (agarose or polyacrylamide gel) and a voltage is applied.
  - o The DNA is loaded on the gel side with the negative current because DNA is negatively charged and will tend to migrate toward the positive current.
  - o Smaller pieces of DNA can move through the gel pores more efficiently and more quickly and will migrate through the gel much quicker.
  - o When the voltage is stopped, the smallest bands will have traversed the farthest distance through the gel medium. The largest DNA bands will remain close to the starting well as their migration was slowed due to the large size.
  - o The distances that the bands migrate through the gel can be used to isolate certain DNA fragments and be used for DNA typing for forensics, paternity, or identification.

Lab #7: Genetics of Organisms

- *Physical characteristics are passed from generation to generation by differing modes of inheritance and involve analyzing parent and offspring characteristics and ratios.*
  - o The **genotype** of an organism is its genetic makeup, while the phenotype describes the physical appearance of an organism.

- o **Monohybrid crosses** involve a single contrasting pair of characteristics.
- o **Dihybrid crosses** involve two pairs of contrasting traits, considered simultaneously.
- o **Sex-linked crosses** involve inheritance associated with the sex chromosomes (usually the x).
- o During meiosis I there can be an exchange of genetic material between non-sister chromatids. This process increases genetic variability of offspring and potentially reproductive success.
- o Often times the results of an experiment do not match the predicted results exactly. In order to determine whether or not this variance is significant, statistical analysis, such as a chi-square test may be used.
- o The formula for chi-square is:

$$X^2 = \text{the sum of } \frac{(o-e)^2}{e}$$

where: o = observed number of individuals
e = expected number of individuals

This value is then compared to the values in the following table. The degrees of freedom (df) value is the number of phenotypic classes minus 1. Under the degrees of freedom column, find your critical value. If the calculated value for chi-square is greater than or equal than the critical value in the table, then you must reject your null hypothesis (there is reason to doubt the proposed hypothesis). The probability indicates what percentage of the time the results are expected to be similar.

Degrees of Freedom

Probability (p)	1	2	3	4	5
0.05	3.84	5.99	7.82	9.49	11.1
0.01	6.64	9.21	11.3	13.2	15.1
0.001	10.8	13.8	16.3	18.5	20.5

Lab #8: Population Genetics and Evolution

- • *The Hardy-Weinberg equilibrium provides a mathematical way to study the allele frequency changes within a population.*
    - o If the following Hardy-Weinberg conditions are maintained, the population's allele and genotype frequencies will remain constant:

- Large breeding population
- Random mating
- No mutation of alleles
- No differential migration
- No selection

- If A and a are alleles for a gene and each individual (diploid) carries two alleles, then p is the frequency of the A (dominant) allele and q is the frequency of the a (recessive) allele.
- Populations in genetic equilibrium are represented by the following equations:

$$p + q = 1.0 \ (100\%)$$

$$p^2 + 2pq + q^2 = 1$$

where: p^2 = frequency of the homozygous dominant genotype
$2pq$ = frequency of the heterozygote genotype
q^2 = frequency of homozygote recessive genotype

Lab #9: Transpiration

- *Plants must find a compromise between an efficient photosynthetic rate and minimal water loss; this is influenced by temperature, humidity, wind conditions and light availability. Experimentation is used to investigate this compromise.*
  - **Transpiration** is specifically the evaporation of water from the surface of a leaf. It is the driving force in water movement through plants.
  - **Hydrogen bonding** between water molecules influences **cohesion** and **adhesion** of water and greatly impacts water movement.
    - **Cohesion** is the attraction of like molecules due to hydrogen bonds.
    - **Adhesion** is the attraction of unlike molecules to each other.
  - **Stomata** are pores in the lower epidermis of the leaf that allow gas exchange between the environment and the plant.
  - **Guard cells** surround the stomata and regulate their opening and closing, which impacts photosynthesis and water concentration.

Lab #10: Physiology of the Circulatory System

- *Heart rate varies with a change in body position, physical fitness of the individual, and physiological habits.*
  - Blood pressure is the measure of the hydrostatic force blood exerts on vessel walls. The two components of blood pressure are **systolic** pressure (ventricular contraction) and **diastolic pressure** (ventricular relaxation).
  - A sphygmomanometer is used to measure blood pressure.
  - An increase in cellular respiration requires more oxygen availability to cells, thus increasing breathing and heart rates.
  - Both endotherms and ectotherms must use **thermoregulatory mechanisms** to maintain internal temperatures.

- Endotherms: panting, sweating, etc.
- Ectotherms: basking, mud bathing, etc.

Lab #11: Animal Behavior

- *Behavior is either <u>learned</u> or <u>innate</u>. By investigating behavior and environmental conditions, an organism's response to its environmental sensory input may be determined.*
  - Organisms behave in ways to put themselves in the most beneficial environment possible. This behavior may be directed towards a stimulus (**taxis**) or random (**kinesis**).
  - Organisms often behave in different ways towards others of the same species or different species (**agonistic behavior**). This behavior may be aggressive or submissive.
  - Mating behaviors are very complex behaviors, especially in *Drosophila melanogaster*. Five phases are present in the mating ritual:
    1. orientation
    2. male song
    3. licking
    4. attempted copulation
    5. copulation or rejection

Lab #12: Dissolved Oxygen and Aquatic Primary Productivity

- *Oxygen enters the water by the process of diffusion from the surrounding air and photosynthesis by aquatic plants. This <u>dissolved oxygen</u> concentration is critical to the survival of aquatic organisms and can be measured in the laboratory.*
  - Specific factors affect dissolved oxygen concentration including: salinity, pH and temperature.
  - Ecosystems store organic compounds at a specific rate. This primary productivity can be measured by analyzing carbon dioxide consumption, synthesis of organic compounds, or oxygen synthesis.
  - When photosynthetic organisms are exposed to light, photosynthesis and respiration are taking place. By analyzing both of these processes, we can determine **net productivity**.
  - When photosynthetic organisms are not exposed to light, no photosynthesis can occur and the only change in **primary productivity** can be attributed to respiration.
  - Autotrophs are able to use sunlight to create simple organic molecules from water and carbon dioxide.

$$6CO_2 + 6H_2O \xrightarrow[\text{sunlight}]{\text{chlorophyll}} C_6H_{12}O_6 + 6O_2$$

MULTIPLE CHOICE QUESTIONS

1. Which population in the Hardy-Weinberg equilibrium will become increasingly smaller if homozygotes selectively mate with individuals of the same genotype?

   (A) p^2
   (B) q^2
   (C) 2pq
   (D) p
   (E) q

2. In *Drosophila*, a cross of a white eyed male with a heterozygous red eyed female, will yield the following results in the F_2 generation:

   (A) 3:1 red eyed males to white eyed females
   (B) 3:1 white eyed males to red eyed females
   (C) 1:1:1:1 red eyed males to white eyed males to red eyed females to white eyed females
   (D) 9:3:4 red eyed females to red eyed males to white eyed males
   (E) 1:2:1 red eyed males to white eyed males to white eyed females

3. Placing a plant in a drought like environment will promote all of the following events EXCEPT?

   (A) an increase in guard cell CO_2 content
   (B) stomatal closure
   (C) increase in turgor pressure of the guard cells
   (D) influx of K^+ into the guard cells
   (E) an increase in water potential of the guard cells

4. Which of the following behaviors is an example of kinesis?

   (A) Organisms responding to bright light by moving away
   (B) Organisms being attracted to a moist environment
   (C) Organisms moving towards a mating sound
   (D) Organisms following a glimmer of light
   (E) Organisms moving in all directions when exposed to light

5. Which of the following patterns is NOT an example of a male *Drosophila* courtship behavior?

   (A)   wing vibration
   (B)   waving
   (C)   ignoring
   (D)   tapping
   (E)   circling

6. The solubility of oxygen in water decreases as

   (A)   temperature increases
   (B)   pH neutralizes
   (C)   temperature decreases
   (D)   salinity decreases
   (E)   turbidity decreases

7. Using the Light-Dark Bottle Method, gross productivity is a measure of:

   (A)   Light – Initial Bottle
   (B)   Initial – Dark Bottle
   (C)   Initial + Light Bottle
   (D)   Light – Dark Bottle
   (E)   Initial + Dark Bottle

8. Which of the following change(s) in heart rate and blood pressure would be expected when a person moves from a lying to a standing position.

   (A)   Blood pressure increase and heart rate decrease
   (B)   Blood pressure decrease and heart rate decrease
   (C)   Blood pressure decrease and heart rate increase
   (D)   Blood pressure increase and heart rate increase
   (E)   Blood pressure increases and heart rate remains steady

9. Which of the following is NOT an adaptation to reduce leaf water loss?

    (A)   reduced number of stomates
    (B)   decrease in leaf surface area
    (C) C_4 photosynthesis
    (D)   sunken stomates
    (E) C_3 photosynthesis

10. Ectotherms increase body temperature through all of the following means EXCEPT

    (A)   basking
    (B)   burrowing
    (C)   changing wing angle
    (D)   changing posture
    (E)   changing body angle

11. Charles Darwin is responsible for helping us to understand all of the following information about descent EXCEPT that

    (A)   organisms experience gradual change due to environmental pressure.
    (B)   chromosomal inheritance is the fundamental mechanism for the passing of traits from parent to off-spring.
    (C)   traits in a population vary.
    (D)   organisms with the traits that are most-suitable for a given environment have the greatest reproductive fitness in that environment.
    (E)   there is competition for limited resources.

12. James Watson and Francis Crick are most noted for the DNA double-helix model that they proposed in 1953. What researchers did not provide key pieces of evidence about the function and structure of DNA that helped Watson and Crick reach this ground-breaking discovery?

    (A)   Rosalind Franklin and Maurice Wilkins
    (B)   Erwin Chargaff
    (C)   Matthew Meselson and Franklin Stahl
    (D)   Alfred Hershey and Martha Chase
    (E)   Frederick Griffith

13. Gregor Mendel did extensive research with garden peas. Through Mendel's experiments with peas, all of the following about inheritance was discerned EXCEPT that

   (A) some variations of traits are dominant and are always expressed when present.
   (B) incomplete dominance sometimes occurs resulting in a heterozygous trait that is intermediate to the two parental traits.
   (C) recessive traits only appear if two recessive "characters" are present for that trait.
   (D) the two alleles that are present for any given trait assort independently from the alleles for other traits during gamete formation.
   (E) each individual has two alleles or "characters" for each trait.

14. If a semi-permeable dialysis bag containing distilled water was placed into a solution of 20 % sucrose

   (A) there would be net movement of water molecules into the bag.
   (B) there would be net movement of water molecules out of the bag.
   (C) there would be net movement of sucrose molecules into the bag.
   (D) there would be net movement of sucrose molecules out of the bag.
   (E) no water molecules or sucrose molecules would move across the dialysis membrane.

15. If a piece of semi-permeable dialysis tubing containing distilled water was placed into a beaker of distilled water,

   (A) there would be a net movement of water molecules into the tubing.
   (B) there would be a net movement of water molecules out of the tubing.
   (C) no water molecules would move across the tubing.
   (D) there would be an equal amount of water molecules moving into the tubing as moving out of the tubing.
   (E) the tubing would burst as water quickly rushed into it.

16. If someone wanted to measure the rate of cellular respiration by measuring the amount of oxygen gas that is consumed in a given environment, it would be best to

   (A) use a probe that measures overall gas pressure in the container.
   (B) compare the gas pressure inside in the enclosed environment to the outside of the enclosed environment.
   (C) use potassium hydroxide to absorb the carbon dioxide that is produced so that the oxygen gas concentration can be more effectively measured.
   (D) use a liquid environment to measure bubbling changes.
   (E) the rate of oxygen consumption cannot be accurately assessed due to confounding factors in the reaction.

17. Mitosis and meiosis are both forms of nuclear division. When the two processes are compared in animals, all of the following would be true of meiosis but **NOT** mitosis EXCEPT

(A) chromosome number of the daughter cells is half the parent cell number.
(B) daughter cells are genetically unique.
(C) individual chromosomes segregate randomly.
(D) the product cells are gametes.
(E) crossing over occurs.

18. Restriction enzymes are important for electrophoresis because

(A) they cut large DNA pieces into smaller fragments that can migrate through the gel.
(B) they unwind the DNA double helix into separate strands.
(C) they amplify small DNA samples into larger volumes for more accurate comparison of banding patterns.
(D) they dye the DNA or cause it to fluoresce to increase the visibility of the DNA bands.
(E) they prevent bacteria from growing on the gel surface.

19. During the transformation of bacteria with engineered plasmids, the bacteria are often stressed in order to

(A) prevent the bacteria from dividing too rapidly and contaminating other samples.
(B) increase the rate of binary fission to increase the number of bacterial cells.
(C) increase the likelihood that the bacteria will accept the engineered plasmids into its interior.
(D) build a protective shield to prevent other plasmids from entering the bacteria.
(E) prevent conjugation between neighboring bacterial cells.

20. Photosynthesis is performed by

(A) autotrophs only.
(B) heterotrophs only.
(C) both autotrophs and heterotrophs.
(D) all living organisms.
(E) none of these answers correctly complete the statement.

FREE RESPONSE QUESTIONS

1. Enzymes are proteins that are extremely important to the efficient and accurate completion of biological reactions.

   a. **Explain** how enzymes increase the rate of biological reactions.
   b. **Explain** how the relationship between enzyme and its triggered reaction are specific.
   c. **Explain** how the following factors can affect the rate of an enzyme-catalyzed reaction: temperature, pH, noncompetitive inhibitors, and competitive inhibitors.
   d. **Design** a brief experiment to measure how temperature would affect the rate of an enzyme catalyzed reaction.

2. Transpiration is the driving force in water absorption of plants. Plants depend on many physiological adaptations and chemical properties of water in order to survive.

   a. **Explain** the structure and function of the leaf to water regulation.
   b. **Discuss** the nature of water and how its unique chemical properties impact water regulation in plants.
   c. Propose and **explain** three adaptations used by plants to overcome difficulties in water regulation, especially by those plants in arid environments.

NO TESTING MATERIAL PRINTED ON THIS PAGE

GO ON TO THE NEXT PAGE

Theme Two

Evolution

- *Biological classification provides a means of grouping organisms that share physiological and structural similarities. This evidence helps to develop hypotheses about evolutionary relationships.*
  - Carolus Linnaeus was the founder of the traditional classification scheme. The classification system was built upon a system of more-inclusive taxa beginning with the species as the most exclusive.
  - Linnaeus used **binomial nomenclature** to identify each organism. This method uses the genus and species of each organism to form a scientific name that is common in all cultures.
  - This classification system has been modified for over 100 years. The most recent modification added a new taxon: the domain.
- *The domain was added because in previous systems, all prokaryotic organisms were placed together in one kingdom. With advances in molecular techniques, it was found that true bacteria and archae are very diverse.*
  - These differences were so great that it warranted the separation of that one kingdom into two different domains: Bacteria and Archae. Members of all other kingdoms have more molecular similarities than members of these two domains. So, all other kingdoms were placed into a common domain: Eukarya
  - Modern Classification System from most inclusive to most exclusive.
    - Domain
    - Kingdom
    - Phylum
    - Class
    - Family
    - Genus
    - Species

17

Three Domains of Life

Domain Name	Features of Domain Members
Domain Bacteria	Prokaryotic cells, diverse life modes (can be nitrogen-fixing, photosynthetic, parasitic), often have cell walls of peptidoglycan, most use oxygen gas for cellular respiration, can be autotrophic or heterotrophic, unicellular, often have flagella
Domain Archae	Prokaryotic cells, live in extreme environments like salt marshes and deep sea vents, often live in anaerobic conditions, unicellular
Domain Eukarya	Eukaryotic cells, includes unicellular and multicellular forms, includes autotrophs and heterotrophs, includes four major kingdoms of organisms (Protista, Plantae, Animalia, Fungi)

Four Kingdoms of Domain Eukarya

Kingdom Name	Features of Kingdom Members
Protista	Unicellular, can be autotrophic or heterotrophic, complex organelle structure, can be free-living or parasitic, can have flagella or cilia
Plantae	Multicellular, terrestrial, autotrophic, cell walls of cellulose, chlorophyll a as the primary photosynthetic pigment, alternating life cycle
Fungi	Multicellular, heterotrophic, cell walls of chitin, decomposers, mycelia form following plasmogamy, haploid life cycle
Animalia	Multicellular, motile, no cell walls, heterotrophic, most reproduce sexually, diploid life cycle

Major Groups of Clades in Kingdom Protista

Group Name	Prominent Features
Diplomonads and Parabasalids	Ancient phyla with modified mitochondria, mitochondria lack DNA and materials for cellular respiration, *Giardia* is a common diplomonad, *Trichomonas vaginalis* is a common parabasalid that causes a common STD
Euglenazoa	Prominent flagella, can be free-livng or parasitic, can be autotrophic or heterotrophic, crystalline rod inside the flagella, examples are the photosynthetic *Euglena* and the parasitic *Trypanosoma* that causes sleeping sickness
Alveolata	Membrane-bound sac called alveoli under membrane, sac may help to regulate water and solute concentrations, dinoflagellates are heterotrophic plankton, apicomplexans are parasitic like *Plasmodium* that causes malaria, ciliates are a large group of protists with cilia surrounding the entire cell membrane for locomotion and two types of nuclei

Stramenopila	Flagella with tiny hairlike projections, can be heterotrophic or autotrophic, the oomycetes are white rusts with cell walls of cellulose, the diatoms have silica walls and are photosynthetic, the chrysophytes are golden algae and are photosynthetic, the phaephytes are brown algae and have very large bodies like kelp
Foraminiferans and Radiolarians	Threadlike pseudopodia that extend through tiny pores of covering, foraminiferans have porous shells of calcium carbonate, radiolarians have outer coverings made of silica
Amoebas	Pseudopodia (lobe-shaped or threadlike), live in soil, freshwater, and marine environments, use pseudopodia for movement and for phagocytosis
Slime Molds	Plasmodial slime molds contain a large plasmodium (single –celled mass of many nuclei in a slime material), Cellular slime molds have separate cells that can form a unit when nutrients are depleted
Rhodophytes	Red algae with pigment phycoerythrin, many are multicellular, alternation of generation life cycle

Major Plant Phyla

Plant Group	Characteristic Features
Bryophytes	Flat body styles, no vascular tissue, no seeds, no fruit, no flowers, sperm must travel through water to reach egg, includes mosses, liverworts, and hornworts
Pterophytes	Vascular tissue (xylem and phloem), axial body style (tall and lean), lacks seeds, produce spores, sperm travel in water to reach eggs, includes ferns, club mosses, and horsetails
Gymnosperms	Vascular tissue, axial body style, seeds (naked seeds that are housed in cones), sperm are airborne and travel by wind to reach eggs that are maintained in ovulate cones
Angiosperms	Vascular tissue, axial body style, seeds that are protected by seed coat, fruit grows from ovary walls to attract animals to disperse seeds, flowers present with advanced reproductive features, double fertilization may occur

Major Fungal Phyla

Phylum Name	Primary Features of Phylum
Zygomycota	Fast growing molds that rot fruit, animal parasites and black bread mold, coenocytic hyphae
Basidiomycota	Mushrooms and shelf fungi, important decomposers of wood, mostly sexual reproduction, fruiting bodies called basidiocarps
Ascomycota	Sac like spores, fruiting bodies called ascocarps, includes some unicellular yeasts and many multicellular forms, sexual and asexual reproduction
Deuteromycota	Imperfect fungi, asexual reproduction, many are unicellular

- *Similarities in animal features provide for variation among species and similarities within phyla.*

Feature	Feature Description and Variations
Symmetry	Radial symmetry is when an organism can be divided into equal halves at multiple locations, bilateral symmetry occurs when organisms can only be cut at one plane to form equal halves
Tissues	Parazoa lack true tissues, eumatzoa have cells organized into tissue layers, diploblastic animals have only two tissue layers (endoderm and ectoderm) during development, triploblastic animals have endoderm, ectoderm, and mesoderm
Body Cavity	Acoelomates lack a body cavity between the three developmental layers, pseudocoelomates have a body cavity between the mesoderm and endoderm, coelomates have a body cavity completely lined with mesoderm tissue
Developmental Path	In protostome development, the blastopore becomes the mouth and cleavage is spiraland determinate; in deuterostomes, the blastopore becomes the anus and the cleavage is radial and indeterminate

• *Evolutionary relationships can be reflected from phylogenetic similarities and differences.*

Phylum	Primary Features of Phylum Members
Porifera	Lack true tissues, no body symmetry, aquatic, food is ingested by using collar cells, both sexual and asexual reproduction, includes sponges
Cnidaria	Have true tissues, radial symmetry as adults and juveniles, two body styles: sessile polyp and motile medusa, includes jellyfish, coral, sea anemone, diploblastic
Platyhelminthes	Flatworms, includes free-living planarians and parasitic flukeworms and tapeworms, bilateral symmetry, triploblastic, acoelomate
Rotifera	Triploblastic, bilateral symmetry, pseudocoelomate, funnel of cilia, reproduce through parthenogenesis
Nematoda	Triploblastic, bilateral symmetry, smooth body, roundworms, pseudocoelomate, very abundant
Nemertea	Ribbon worms, bilateral symmetry, triploblastic, long proboscis for feeding, closed-circulatory system, marine
Lophophorates	Lophophore for filtering food from water, coelomate, includes brachiopods and phoronids
Mollusca	Bilateral symmetry, large muscular foot, organs in visceral mass, mantle tissue, may have calcium shell, open circulatory system, coelomate, clams, snails, squid, and octopus
Annelida	Segmented worms, coelomate, bilateral symmetry, can live in water or on land, earthworms, leeches, and marine worms, closed-circulatory system
Arthropoda	Bilateral symmetry, protostome, coelomate, exoskeleton of chitin, insects, spiders, and crustaceans, many jointed appendages that can be specialized for diverse functions
Echinodermata	Radial symmetry as adults, bilateral symmetry as juveniles, water vascular system, tube feet for attaching to surfaces and to open shells of prey, starfish and sea cucumbers, deuterostome
Chordata	Deuterostome, bilateral symmetry, coelomate, closed-circulatory system, dorsal hollow nerve cord, pharyngeal gill slits, post anal tail, notochord

Class	Primary Features
Agnatha	Jawless fish, hagfish and lampreys, attach to prey body and absorb nutrients
Chondricthyes	Fish with cartilaginous skeleton, sharks and rays, can have live birth after internal fertilization
Osteicthyes	Bony fish, includes ray-finned and lobe-finned fish, most have external fertilization, skeleton of hardened bone, includes tuna, salmon, and sunfish, two-chambered heart
Amphibia	Tetrapods, live on land and water, juveniles are tied to the water, external fertilization in water, jellylike eggs are laid in the water, get oxygen through lungs and across skin, three-chambered heart
Reptilia	Tetrapods with leathery skin, terrestrial, internal fertilization, lay amniote eggs with leathery shells, ectothermic, three-chambered heart (some members have four-chambered heart)
Aves	Endothermic, four-chambered heart, feathers, porous bones for buoyancy, internal fertilization, lay amniote eggs with calcium shells, sometimes included in Class Reptilia
Mammalia	Endothermic, four-chambered heart, internal fertilization, amniote eggs develop inside the maternal body, hair covering body, mammary glands to nourish young

Major Classes in Phylum Chordata

• *Natural selection is a driving force for evolution and may act upon a population in a variety of ways, along with other factors.*
  o Mutation introduces new alleles.
  o Emigration and immigration impact allele frequency.
  o A random increase or decrease of alleles affects populations.
  o Mating patterns, such as inbreeding and selection, affect allele frequency.
  o **Stabilizing** selection works against individuals who exhibit any one of the extreme traits and favors intermediate phenotypes.
  o **Directional selection** favors individuals who exhibit one type of extreme characteristic.
  o **Disruptive selection** works against individuals who possess non-extreme traits.

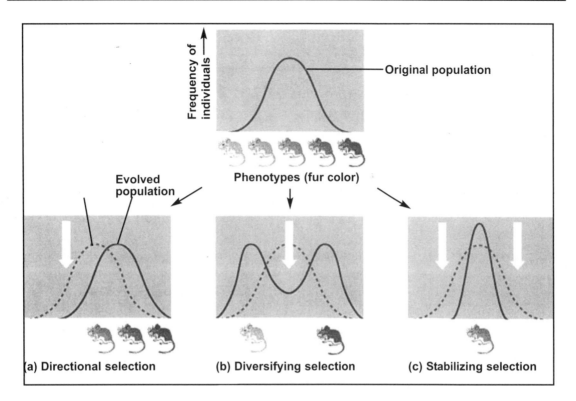

Fig. 23.12, p. 458 from BIOLOGY, 6th ed. by Neil A. Campbell and Jane B. Reece. Copyright © 2002 by Pearson Education, Inc. Reprinted by permission.

- *Survival and reproduction are necessary to affect future generation's changes in allele frequencies.*
  - Ecosystems possess unique carrying capacities that have limited resources, competition, and reproductive potential. Organisms that survive and reproduce will pass on traits to the next generation.
  - Variation is heritable.
  - Evolution occurs as traits accumulate in a population.
- *Variation within a population is a necessary condition for natural selection to occur. This variation occurs through the following processes:*
  - Mutation
  - Crossing over
  - Random assortment
  - Random fertilization
  - Diploidy
- *Evidence for evolution spans several scientific disciplines and helps us to determine evolutionary relationships.*
- *Examination of the amino acid sequences of DNA through molecular biology techniques reveals that closely related species exhibit similar nucleotide sequences.*

o Structural similarities of body parts give rise to the understanding of evolutionary relationships.

■ **Analogous structures:** structures that have evolved a similar form due to similar environment pressures
■ **Vestigial organs:** evolutionary relics that have little or no function in the modern species
■ **Homologous structures**: structures that function differently but evolved from a common structure

• *When variation advances to the point of preventing interbreeding, speciation occurs under specific conditions.*
  o **Allopatric speciation** results when a geographic barrier prohibits interbreeding. These separate populations, although still one species, due to environmental differences or mutation, may diverge to become separate species.
  o **Sympatric speciation** does not require a geographic barrier. However, balanced polymorphism, polyploidy, and hybridization may cause the formation of a new species.
  o **Adaptive radiation** is a type of colonization where many species diverge from a single ancestor.
• *The Hardy-Weinberg equilibrium provides a mathematical way to study the allele frequency changes within a population.*
  o If the following Hardy-Weinberg conditions are maintained, the population's allele and genotype frequencies will remain constant:
    ■ Large breeding population
    ■ Random mating
    ■ No mutation of alleles
    ■ No differential migration
    ■ No selection

  o If A and a are alleles for a gene and each individual (diploid) carries two alleles, then p is the frequency of the A (dominant) allele and q is the frequency of the a (recessive) allele.
  o Populations in genetic equilibrium are represented by the following equations:

$$p + q = 1.0 \ (100\%)$$

$$p^2 + 2pq + q^2 = 1$$

where: p^2 = frequency of the homozygous dominant genotype
$2pq$ = frequency of the heterozygote genotype
q^2 = frequency of homozygote recessive genotype

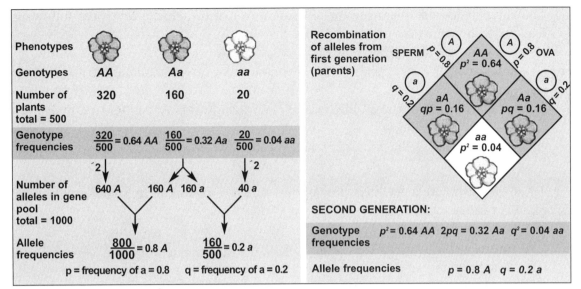

(a) Genetic Structure of parent population   (b) Genetic Structure of second generation

Fig. 23.3, p. 448 from BIOLOGY, 6th ed. by Neil A. Campbell and Jane B. Reece. Copyright © 2002 by Pearson Education, Inc. Reprinted by permission.

MULTIPLE CHOICE QUESTIONS

1. The charophytes, an algal ancestor to modern-day plants, shared all of the following features with the modern plants EXCEPT for

    (A)   photosynthetic autotroph
    (B)   lived on land
    (C)   alternation of generation life cycle
    (D)   cell walls of cellulose
    (E)   chlorophyll a as the primary photosynthetic pigment

2. These organisms are unicellular, prokaryotes who live in extreme environments and are often anaerobic

    (A)   viruses
    (B)   true bacteria
    (C)   blue-green algae
    (D)   archae
    (E)   prions

3. The Domain taxon was added to the traditional classification system because

    (A)   new species of organisms were discovered
    (B)   Archae and bacteria are very dissimilar genetically
    (C)   there is a great diversity of eukaryotic organisms
    (D)   there wasn't any place for viruses
    (E)   the increase in species number required a new taxon

4. Organisms with prokaryotic cells are classified in the Domain

    (A)   Bacteria
    (B)   Archae
    (C)   Eukarya
    (D)   both A and B
    (E)   A, B, and C

5. Multicellular, terrestrial autotrophs with cell walls of cellulose and alternating life cycles belong to the Kingdom

    (A)   Animalia
    (B)   Protista
    (C)   Monera
    (D)   Plantae
    (E)   Fungi

6. Adult, radial symmetry is seen in

    (A)   flatworms of Phylum Platyhelminthes
    (B)   jellyfish and Hydra of Phylum Cnidaria
    (C)   shellfish of Phylum Arthropoda
    (D)   starfish of Phylum Echinodermata
    (E)   both the Echinoderms and the Cnidarians

7. As a ciliated animal-like protist, *Bursaria* is likely to be most closely related to which of the following protists?

    (A)   plasma-like amoeba
    (B)   shelled foraminiferan
    (C)   cell-walled chlorophyte
    (D)   motile, fresh-water paramecium
    (E)   slime mold

For questions 8-10, use the following answer choices.

   (A)   Class Mammalia
   (B)   Class Reptilia
   (C)   Class Aves
   (D)   Class Arthropoda
   (E)   Class Chondricthyes

8. The first mammals evolved from members of this class.

9. The first birds evolved from members of this class.

10. Amniote eggs are found in

   (A)   A only
   (B)   D only
   (C)   B and C
   (D)   A, B, and C
   (E)   A, B, C, and D

For questions 11-15, use the following answer choices.

   (A)   directional selection
   (B)   stabilizing selection
   (C)   disruptive selection
   (D)   standardizing selection
   (E)   sexual selection

11. Light peppered moths are less common in areas of high pollution than dark peppered moths.

12. Dandelions tend to be shorter in lawns than they do in the wild.

13. Two different beak sizes, small billed and large billed, occur in a single population of finch.

14. Female robins who lay three to five eggs at a time have more surviving young than those who lay more or less eggs.

15. Female fiddler crabs prefer males with larger claws.

16. Which of the following does NOT promote variation within a population?

   (A) crossing over
   (B) mutation
   (C) independent assortment
   (D) inbreeding
   (E) random fertilization

17. A hurricane devastates an area of the southern United States, causing mass extinction of several species. The remaining organisms quickly diversify and begin establishing themselves in the new ecosystem, presenting new opportunities and problems. This is an example of

   (A) founder effect
   (B) sympatric speciation
   (C) allopatric speciation
   (D) genetic drift
   (E) adaptive radiation

18. An increase in the chromosomal complement of a species, greater than 2n, results in

   (A) Diploidy
   (B) Polyploidy
   (C) Haploidy
   (D) Gametoploidy
   (E) Aneuploidy

19. A given population is in Hardy-Weinberg equilibrium. The frequency of the recessive phenotype is symbolized by

   (A) $p^2 + 2pq$
   (B) either p or q
   (C) p^2
   (D) q^2
   (E) q

20. Variation within a population occurs because of several different mechanisms. Which of the following choices promotes variation during meiosis I due to the exchange of genetic material between homologous chromosomes?

   (A) mutation
   (B) crossing over
   (C) random assortment
   (D) random fertilization
   (E) polyploidy

FREE REPONSE QUESTIONS

1. Use the following features to describe the differences between these four groups of land plants: Bryophytes, Pterophytes, Coniferophytes (Gymnosperms), and Anthophytes (Angiosperms). **Explain** what each feature is and **describe** how it is involved in determining classification.

    a.   Vascular Tissue
    b.   Seeds
    c.   Flowers
    d.   Fruits

2. The Hardy Weinberg theorem describes a nonevolving population, a gene pool that remains constant over time. For the following population:

In a population with two alleles for a particular locus, A and a, the allele frequency of A is 0.7. Of the 20,000 turtles in the population, approximately 1,800 express the recessive phenotype for this particular gene.

    a.   **Identify** whether or not the population is in Hardy Weinberg equilibrium and calculate the allele and genotypic frequencies in the population's gene pool.
    b.   **Identify** three of the main conditions that must be present in order for a population to be in Hardy Weinberg equilibrium.
    c.   **Describe** three of the main factors that can act to alter gene frequencies and cause microevolution to occur.

Theme Three

Energy Transfer

FERMENTATION

- *There are two primary cellular energy-producing processes: fermentation and cellular respiration. In these processes, chemical energy in the form of glucose is converted to ATP, the primary form of cellular energy.*
  - o Fermentation is much less efficient than cellular respiration—the process nets only two ATP molecules per glucose molecule invested in the process.
  - o There are two primary steps to fermentation: the ATP-producing step of glycolysis and the following fermentation step to reform NAD+.
  - o The process is anaerobic and occurs in cells when no oxygen gas is present.
  - o All stages occur in the cytosol of the cell.

GLYCOLYSIS

  - o Glycolysis is the oldest energy forming process. It occurs in almost every living organism and has been conserved throughout the process of evolution.
  - o One glucose molecule is used to begin the process.
  - o There are two stages: the energy investment stage and the energy payoff phase.
  - o In the energy investment phase, the energy of two ATP molecules is used to begin the process.
  - o The released phosphates from breaking down the two ATPs are transferred to the glucose to make it unstable. It quickly divides into two 3-carbon sugars: glyceraldeyde-3-phosphate.
  - o In the energy payoff phase, the glyceraldehydes-3-phosphates are reconfigured to form ATP molecules and pyruvates.
  - o The phosphates are removed from the two glyceraldehyde-3-phosphates and the atoms are reconfigured to form two 3-carbon pyruvates.
  - o During the process, two pairs of electrons are transferred to two NAD+ (an electron carrier) molecules to make two NADH.
  - o The reconfiguration of each 3-carbon molecule provides enough energy to make two ATP molecules (four total).

FERMENTATION STAGE

  - o During this stage, **no** more ATP are produced. The goal of this stage is to convert the NADH back to NAD+ so that more glycolysis can occur.
  - o There are two primary types: **alcohol fermentation** that occurs commonly in plants, yeast, and bacteria and **lactic acid fermentation** that occurs commonly in animals.

31

o In alcohol fermentation, each of the NADH molecules produced in glycolysis transfers a pair of electrons to a pyruvate molecule, which makes it unstable. The pyruvate breaks into a carbon dioxide and a two-carbon ethanol. The NADH molecules become NAD+ which can return to glycolysis.

o In lactic acid fermentation, each of the NADH molecules produced in glycolysis transfers a pair of electrons to a pyruvate molecule, which makes it unstable. The pyruvate reconfigures into a 3-carbon product, lactic acid. The NADH molecules become NAD+ which can return to glycolysis.

o Net products are two ATP, two NADH, and two pyruvates.

CELLULAR RESPIRATION

o Cellular respiration begins with the previously-described process of glycolysis.

o The pyruvates are moved to the interior of mitochondria where the process is continued.

o There are four distinct stages: **glycolysis**, **pyruvate oxidation**, the **Krebs cycle**, **oxidative phosphorylation**.

o The process is very efficient, converting one glucose into a net of 36 to 38 ATP molecules.

o The process is oxygen dependent. Oxygen gas is required for the oxidative phosphorylation stage. No other stage of cellular respiration will follow glycolysis if no oxygen gas is available.

o The process requires glucose and oxygen to produce water and carbon dioxide and ATP.

PYRUVATE OXIDATION

o In this process, the two pyruvates from glycolysis are moved to the mitochondrial matrix by active transport for the Krebs cycle.

o The two pyruvates are oxidized as each passes a pair of electrons to an NAD+ molecule to yield two NADH molecules.

o As the electrons are lost, each 3-carbon molecule forms a carbon dioxide and a two-carbon acetyl group that is bound to Coenzyme A.

o Net products are two NADH, two CO_2, and two acetyl CoA.

KREBS CYCLE

o The Krebs Cycle occurs in the mitochondrial matrix.

o Each acetyl CoA enters the cycle producing two times the described products.

o The CoA breaks off and attaches the acetyl group to a four-carbon oxaloacetate to produce a six-carbon citrate.

o This 6-carbon molecule is reconfigured and a carbon is broken off as CO_2. A NADH is produced from NAD+ in the process.

o The remaining 5-carbon molecule experiences another reconfiguration and carbon loss producing another CO_2 and another NADH.

- The remaining 4-carbon molecule experiences several reconfigurations that yield one GTP that is converted to ATP, one more NADH, and one $FADH_2$ (another electron carrier).
- Then, the 4-carbon oxaloacetate is regenerated so that the cycle can happen again.
- Net products from two aceytl CoA molecules include four CO_2, six NADH, two $FADH_2$, and two ATP

OXIDATIVE PHOSPHORYLATION

- In this process, the NADH and $FADH_2$ molecules transfer electrons to a series of electronegative proteins located in the inner mitochondrial membrane.
- This inner membrane has many folds called **cristae** that dramatically increase the surface area available for this process.
- The $FADH_2$ molecule is more electronegative than NADH. So, it transfers its electrons later in the chain that results in fewer passes and less efficient transfer of energy. So, each $FADH_2$ eventually yields two ATP molecules in this process. Each NADH yields three ATP molecules in this process.
- Up to 34 ATP can be made through this process.
- After the electron pairs are transferred from the NADH and $FADH_2$ to the electronegative proteins, they move through a series of proteins. As they are passed energy is released that is used to actively transport H+ ions from the matrix to the intermembrane space.
- At each pass of electrons, the energy release is used to build this proton gradient as the H+ concentration builds in the intermembrane space.
- As the proton gradient is established, the protons move to reach equilibrium by diffusing through a protein, **ATP synthase**.
- ATP synthase is an enzyme that can harness the energy from the H+ diffusion to add a phosphate to ADP and make ATP through **chemiosmosis**.

PHOTOSYNTHESIS

- *This process that occurs in autotrophs is the primary means to make chemical energy in the form of glucose through the transformation of light energy.*
    - This process occurs in specialized areas of the chloroplasts: the light-dependent stages occur in the **thylakoid membranes** and the light-independent processes occur in the **stroma**.
    - There are two major light dependent processes: **cyclic and noncyclic photophosphorylation**, and one major light independent process: **Calvin-Benson cycle**.
    - Photosynthesis requires water, light energy, and carbon dioxide and produces glucose and oxygen gas.

NONCYCLIC PHOTOPHOSPHORYLATION

o In this process, both photosystems (I and II) of the membrane are activated. The photosystems contain pigment molecules that absorb light energy. This energy is transferred to the chlorophyll a of the reaction center where a pair of electrons becomes energized.

o The electrons become energized at photosystem II.

o High-energy electrons from chlorophyll a will be passed through a series of electronegative proteins releasing energy that is used for the active transport of H+ ions from the stroma to the thylakoid lumen.

o The lost pair of electrons is replaced by breaking a water molecule. Oxygen gas and H+ ions are produced by this electron extraction.

o When the electrons reach photosystem I, they are re-energized in a similar way.

o The electrons are again passed through a series of electronegative proteins increasing the H+ gradient between the stroma and the thylakoid lumen.

o Low energy electrons end at an enzyme, NADP reductase that adds the electron pair and a proton to NADP+ to make NADPH.

o Protons diffuse through the ATP synthase enzyme as they move down a concentration gradient. They diffuse from the lumen to the stroma.

o As the H+ ions diffuse, a phosphate is added to ADP to make ATP.

o This process produces O_2, NADPH, and ATP.

CYCLIC PHOTOPHOSPHORYLATION

o Electrons are energized at photosystem I and passed through a series of electronegative proteins. As they are passed, energy is released that is used for the active transport of H+ ions from the stroma to the thylakoid lumen.

o When the electrons return to their low energy state, they return to the chlorophyll in photosystem I and are reused in the process.

o The H+ gradient causes H+ to diffuse from the thylakoid lumen back to the stroma through the ATP synthase enzyme.

o As the diffusion happens, the energy is harnessed to transfer a phosphate to ADP to make ATP.

o Only ATP is made through this process.

CALVIN BENSON CYCLE

o This light-independent process uses the ATP and NADPH from the light-dependent processes as the primary energy source.

o Six cycles of this process must happen to make one glucose molecule.

o In carbon fixation, CO_2 molecules are bound to a five-carbon RuBp by the enzyme **rubisco** to make an unstable 6-carbon compound.

o In sugar production, this 6-carbon divides to form stable 3-carbon glceraldehyde-3-phosphates. For every six cycles, two glyceraldehyde-3-phosphates leave the cycles as products.

○ In regeneration, the other ten glyceraldehyde-3-phosphates from the six cycles are reconfigured to reform the starting RuBp molecules.
○ Following the cycle, the two glcyceraldehyde-3-phosphates are combined to make one glucose

C₄ PLANTS

○ C_4 plants have evolved a mechanism to avoid **photorespiration**, an inefficient process that occurs in plants where oxygen gas is bound in place of carbon dioxide by rubisco in the Calvin-Benson cycle. This photorespiration occurs in C_4 plants when the concentration of CO_2 is reduced.
○ In C_4 plants like pineapples, an enzyme PEP will bind only CO_2 even at low concentrations. The PEP binds the CO_2 and takes it to the rubisco to avoid O_2 from binding at times when stomata are closed and CO_2 concentration is reduced.

CAM PLANTS

○ CAM plants like cacti close stomata during the intense heat of day to prevent water loss.
○ They open stomata at night and convert the CO_2 that enters to organic acid molecules. Then, throughout the day, the acids are turned back to CO_2 so that it is available for the rubisco enzyme in the Calvin-Benson cycle.

ENZYMES

• *Enzymes provide a reduction in energy requirements of reactions so that biological reactions can proceed efficiently.*
  ○ **Enzymes** are proteins with complex three-dimensional shapes.
  ○ Because of this shape, each enzyme is very specific for the type of substrate that it can bind. The enzyme and substrate must have complementary three-dimensional configurations so that the two can bind at the active site.
  ○ Sometimes, there are **cofactors** or **coenzymes** that reinforce the enzyme and substrate connection by binding to the enzyme to make the induced-fit between the enzyme and substrate more distinct.
  ○ The substrate that binds to the enzyme is a reactant from a chemical reaction that will be catalyzed by the binding of the enzyme.
  ○ The enzyme lowers the activation energy of the reaction. The **activation energy** is the energy requirement for the reaction to proceed, and often serves as a barrier to the proceeding of chemical reactions. The enzyme reduces that energy barrier so that the reaction will proceed more quickly.
  ○ After the reaction, the substrate molecule has been consumed and converted to products and the enzyme is released to catalyze another reaction.

o **Competitive inhibitors** are molecules that compete with the substrate for the enzyme's active site. These molecules interfere with the substrate binding to the enzyme by physically blocking access to the active site. The presence of these inhibitors slows the enzyme-catalyzed reaction.

o **Noncompetitive inhibitors** bind to the enzyme outside of the active site. This binding changes the shape of the enzyme and its active site so that it is difficult for the substrate to bind.

o An increase in temperature causes the reaction rate to increase as the molecules move past one another at an increased speed. However, at some point, higher temperatures will cause the **denaturation** of the enzyme which will stop the reaction from occurring.

o Each enzyme has an optimum pH range where it is most efficient. Outside of that range, the enzyme becomes less efficient.

o **Allosteric regulation** involves a molecule binding at a site outside of the active site to make the enzyme more receptive to substrate binding or to make the enzyme inactive by making the active site unavailable.

ANIMAL NUTRITION

• *As a heterotroph, the human body requires the consumption of food for several purposes:*

o FUEL for all cellular activities,

o PRODUCTION of structural building blocks, and

o OBTAINING of essential nutrients.

• *Organisms maintain homeostasis by releasing hormones to allow consumption of fuel in many forms and storage of fuel in another form, transferring energy from one form to another.*

o Organisms store excess glucose as a polysaccharide, **glycogen**. Glycogen is found primarily in two sites in the body – the liver and the skeletal muscles.

o Organisms use the monosaccharide glucose, as fuel, which is converted to ATP to complete cellular work.

o The conversion of glycogen to glucose is regulated by two pancreatic hormones – **insulin** and **glucagon**. As we consume food, glucose and other monomers are absorbed into the bloodstream. If too much glucose is consumed, and the blood glucose level rises to a specific level, a pancreatic hormone, insulin, is released. This hormone causes the liver and muscle cells to increase their permeability to glucose and to convert it to glycogen. As this happens, blood glucose levels drop to an acceptable level. The reverse is true if too little glucose is consumed. In this case, another pancreatic hormone, glucagon, stimulates the conversion of glycogen to glucose that increases blood glucose levels.

INSULIN
(blood glucose levels are too high)

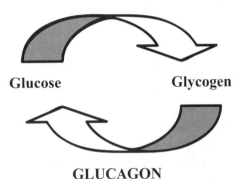

Glucose   Glycogen

GLUCAGON
(blood glucose levels are too low)

- *Feedback systems are integral in controlling energy transfer in organisms.*
  - **Negative Feedback Systems**. Negative feedback systems work like thermostats. As the level of a particular substance rises, there is a limit. Once this limit is reached, a regulatory substance is released in order to bring the level back to homeostasis. If the level goes too low, another regulatory substance is released to cause a rise to reach homeostasis. Examples: blood glucose level regulation (insulin and glycogen) and tempature regulation.
  - **Positive Feedback Systems:** In a positive system, the production of a substance stimulates production of more of that substance. The only time this feedback system ceases to function is if the substance is not needed anymore. Examples: pepsinogen conversion to pepsin, and regulation of bloodclotting.

- *Biosynthesis requires the consumption of many essential nutrients, as the body is unable to make them from any materials. These essential nutrients include: amino acids, fatty acids, vitamins, and minerals.*
  - Many vitamins are used in the body as **coenzymes**.
  - Vitamins can be either fat-soluble or water-soluble. This is a major difference when dealing with human consumption, as water-soluble vitamins, when consumed in moderate excess, are not harmful, as they are simply excreted through the excretory system in the urine. However, fat soluble vitamins, when consumed in excess will be deposited in adipose (fat) cells and accumulate, possibly causing toxic effects to the body.
  - Major water-soluble vitamins and some major diseases due to deficiency or excess: Vitamin B1 (deficiency associated with beriberi), Niacin, Folic Acid, biotin, and Vitamin C (deficiency associated with scurvy).
  - Major fat-soluble vitamins and some major diseases due to deficiency or excess: Vitamin A (excess causes liver and bone damage, vision problems, and intestinal upset), Vitamin D (deficiency associated with rickets), Vitamin E, and Vitamin K (deficiency associated with blood clotting issues).

o Minerals are required for the proper functioning of many body systems, act as cofactors with enzymes, and are the building blocks of hormones. As with vitamins, there may be toxic effects to excess consumption of minerals.

- *Energy transfer occurs within the digestive system through the process of enzymatic hydrolysis. Enzymes break down proteins, fats, and carbohydrates to usable forms and hormones regulate the release of these enzymes.*

o Major digestive enzymes include:

- **Salivary amylase:** This enzyme hydrolyzes polysaccharides in the mouth. It has an optimum pH of 7 and produces smaller polysaccharides and disaccharides

- **Pepsin:** This enzyme hydrolyzes proteins in the stomach, by breaking the peptide bonds that hold polypeptides together. It has an optimum pH of 2. This breakdown is governed by a positive feedback system.

    o **Chief cells** secrete pepsinogen, an inactive precursor to pepsin. This inactive precursor will not be converted to pepsin until HCl is secreted from the **parietal cells** of the stomach.
    o At this point, pepsinogen is cleaved to become an active enzyme, pepsin, which also has the ability to cleave other pepsinogen molecules.

$$\text{Pepsinogen} \xrightarrow{\text{Pepsin or HCl}} \text{Pepsin}$$

- **Pancreatic amylases:** These enzymes complete carbohydrate digestion, forming molecules of glucose that can now be absorbed into the bloodstream.

- **Trypsin, chymotrypsin, carboxypeptidase, aminopeptidase, dipeptidase,** and **enteropeptidase** digest polypeptides in the small intestine, producing dipeptides, which are broken into amino acids as they cross the brush border of the small intestine.

- **Nucleases** break down DNA and RNA in food components into nitrogen containing bases, respective sugars, and phosphates.

- **Bile**, produced by the liver and secreted by the gall bladder, emulsifies fats, increasing surface area and allowing more efficient digestion.

- Fat digestion follows in the small intestine through the enzymatic release of **lipase** after being coated with bile, produced by the liver and secreted by the gall bladder.

o Major digestive hormones include:

- **Gastrin:** the stomach secretes this hormone when food enters the stomach. It stimulates the release of gastric juice (HCl and pepsin). HCl also controls another negative feedback system, only allowing a secretion of gastrin when the pH of the stomach is appropriate.

- **Secretin:** This hormone is secreted by the small intestine when chyme (digested food, gastric juice) enters the duodenum. It stimulates the pancreas to release bicarbonate. This is extremely important as the cells of the small intestine do not have the mechanisms to prevent destruction from an acidic solution, and the enzymes that are present in the small intestine have neutral optimum pHs.

- **Cholecystekinin:** The intestinal wall secretes this hormone at the same time as secretin, when fatty or amino acids are present. This stimulates the pancreas to release pancreatic enzymes and the gallbladder to release bile.

- Other hormones also slow the entry of food into the small intestine, increasing enzymatic efficiency and inhibiting peristalsis.

- *Plants are involved in many symbiotic relationships in order to enhance energy transfer. Without these relationships, plants would not have nitrogen for biosynthesis.*
  - **Nitrogen fixation:** roots of a plant require fixed nitrogen in order for biosynthesis to occur. This is a **mutualistic relationship**, with both the bacteria and the plant benefiting. Fixed nitrogen is supplied to the plant for biosynthesis by the bacteria, and the plant provides carbohydrates to the bacteria.
  - **Mycorrhizae:** a mutualistic relationship where a fungus gains a positive environment to grow and develop, as well as carbohydrates, however provides an increased surface area for water and phosphate absorption for the plants.

- *Plants require essential nutrients (macronutrients and micronutrients) in order to grow from a seed and complete the life cycle.*
  - **Macronutrients:** elements that the plant requires in large amounts. These include carbon, hydrogen, oxygen, nitrogen, sulfur, phosphorous, potassium, calcium and magnesium,
  - **Micronutrients:** elements that the plant requires only in small amounts. These are often used as cofactors with enzymes. Iron helps to play a role in the electron transport chain of chloroplasts and mitochondria as a component of cytochromes.

- *Trophic levels differentiate species in an ecosystem based on their nutritional requirements. These relationships determine the flow of energy within the ecosystem.*
  - **Primary producers:** These organisms are autotrophs (plants, algae, and specific bacteria).
  - **Primary consumers:** These organisms are herbivores and consume primary producers.
  - **Secondary consumers:** These organisms are carnivores and eat primary consumers.

o **Tertiary consumers:** These organisms are carnivores, or omnivores, and eat consumers and/or producers.

o **Decomposers:** These organisms (usually fungi and prokaryotes) break down dead plants and animals.

• *Primary productivity defines the conversion of light energy to chemical energy within an ecosystem. Seasonal effects, such as precipitation, light penetrance and temperature often play a role in the efficiency of this conversion.*

o Ecosystems vary in their productivity. Rainforests are very productive terrestrial ecosystems, as are estuaries. Perhaps the most productive ecosystem is the ocean. However, given the relative size of these ecosystems, rainforests are the most efficient and oceans are the least efficient. The tundra and deserts are also low in productivity.

• *Energy transfer through ecosystems is not 100% efficient.*

o An energy pyramid is used to show the amount of energy at each trophic level. From one level to another, the efficiency is approximately 10%.

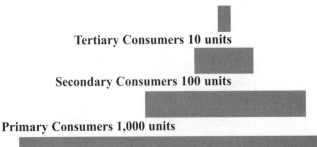

Tertiary Consumers 10 units

Secondary Consumers 100 units

Primary Consumers 1,000 units

Primary Producers 10,000 units

MULTIPLE CHOICE QUESTIONS

1. All of the following organisms perform cellular respiration **except** for

   (A) fungi
   (B) animals
   (C) plants
   (D) protists
   (E) A, B, C, and D

2. ATP synthase is used to make ATP in plants during

   (A) cyclic photosphosphorylation
   (B) noncyclic photosphosphorylation
   (C) chemiosmosis
   (D) both a and b
   (E) A, b and c are correct

3. All of the following reduce the rate of an enzyme-catalyzed reaction EXCEPT

   (A) competitive inhibitors
   (B) noncompetitive inhibitors
   (C) increased enzyme concentration
   (D) extreme pH change
   (E) boiling

4. Examine Figure ET1. Where is the active site of this enzyme?

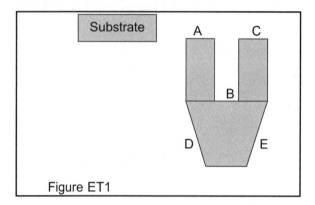

Figure ET1

(A)   A
(B)   B
(C)   C
(D)   D
(E)   E

5. If the enzyme depicted in this figure was subjected to allosteric regulation, which sites on the enzyme surface would be altered when a regulator attached to this enzyme

(A)   A
(B)   B
(C)   C
(D)   D
(E)   A, B, C and D

6. Increasing the temperature of an enzyme-catalyzed reaction from 25° C to 30° C will probably cause

(A)   A slowing of the reaction
(B)   The reaction rate to increase
(C)   The reaction to stop completely
(D)   The reaction to continue at the present rate
(E)   It is impossible to predict how the rate would be affected with this information

7. During the Calvin-Benson cycle in C3 plants, carbon dioxide is bound to a five-carbon ribulose bisphosphate by a special enzyme called

   (A)   phosphoenolpyruvate (PEP)
   (B)   glyceraldehyde 3 Phosphate (G3P)
   (C)   ATP synthase
   (D)   rubisco
   (E)   oxaoloacetic acid

8. During noncyclic photosphosphorylation, all of the following are produced EXCEPT for

   (A)   oxygen gas
   (B)   ATP
   (C)   NADPH
   (D)   carbon dioxide
   (E)   all of these molecules are produced during this process.

9. An energy forming process that is present in almost all living organisms appears to have evolved very early in the evolution of living organisms. This process is

   (A)   krebs cycle
   (B)   glycolysis
   (C)   lactic acid fermentation
   (D)   chemiosmosis
   (E)   cyclic photophosphorylation

10. During the electron transport chain, the final electron acceptor is

   (A)   water
   (B)   oxygen gas
   (C)   carbon dioxide
   (D)   PEP
   (E)   rubisco

11. Which of the following enzymes works to break down polysaccharides?

   (A)   pepsin
   (B)   amylase
   (C)   nuclease
   (D)   trypsin
   (E)   catalase

12. An absence of which of the following substances would result in an inactivation of small intestine enzymes and an erosion of the duodenum?

   (A)   gastrin
   (B)   pepsin
   (C)   secretin
   (D)   cholecystekinin
   (E)   bile

13. Which of the following enzymes works best at a lower pH?

   (A)   salivary amylase
   (B)   pepsin
   (C)   lipase
   (D)   nuclease
   (E)   pepsinogen

14. If 100 units of usable energy are available to the producers in a pyramid of energy, how many units will remain for the secondary consumers?

   (A)   10 units
   (B)   9 units
   (C)   80 units
   (D)   1 unit
   (E)   8 units

15. What are organisms that consume only autotrophs considered to be?

   (A)   heterotrophs
   (B)   omnivores
   (C)   herbivores
   (D)   carnivores
   (E)   detritovores

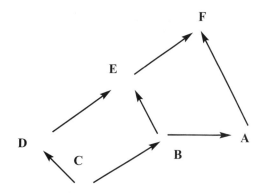

16. The available amount of energy is the least in which species of the following food web

    (A)   A
    (B)   E
    (C)   F
    (D)   D
    (E)   B

17. Which of the following ecosystems is considered to be the most efficient in net productivity?

    (A)   rainforest
    (B)   open ocean
    (C)   desert
    (D)   tundra
    (E)   deciduous forest

18. In the following food chain, which organism is a secondary consumer?

            I. Grass $\longrightarrow$ II. Zebra $\longrightarrow$ III. Lion $\longrightarrow$ IV. Vulture

    (A)   I only
    (B)   II only
    (C)   III only
    (D)   II and III only
    (E)   I, II and III

19. Which hormone functions to cause a release of HCl when food enters the stomach?

    (A) gastrin
    (B) secretin
    (C) cholecystokinin
    (D) bile
    (E) estrogen

20. The process of nitrogen fixation

    (A) converts nitrogen gas to ammonia by prokaryotes
    (B) converts ammonia to nitrates
    (C) converts ammonia to nitrogen gas
    (D) converts nitrogen gas to nitrates
    (E) converts animo acids to urea

FREE RESPONSE QUESTIONS

1. C is cycled through ecosystems geologically and biochemically. CO_2 and O_2 cycle through biochemical processes of living organisms. In a field community, a rabbit population feeds on prairie grass.

   a. **Explain** how carbon is cycled through an ecosystem geologically.
   b. **Explain** how CO_2 and O_2 biochemically cycle through the rabbits and prairie grass in this community. Describe the chemical processes that occur in the rabbits and prairie grass to form these important molecules.
   c. How is the cycling of carbon similar the cycling of nitrogen in an ecosystem?

2. Energy transfer occurs within the digestive system through the process of enzymatic hydrolysis. Enzymes break down proteins, fats, and carbohydrates to usable forms and hormones regulate the release of these enzymes. For this system:

   a. **Describe** how the induced fit theory for enzyme function is of importance in the digestive system.
   b. **Discuss** how pH impacts the efficiency of these enzymes, and explain how molecular interactions influence this process.
   c. **Explain** how feedback systems regulate the release of these enzymes within the digestive system.

NO TESTING MATERIAL PRINTED ON THIS PAGE

GO ON TO THE NEXT PAGE

Theme Four

Continuity and Change

CHROMOSOMAL INHERITANCE

- *The genetic material is contained in chromosomes and inherited from parents to offspring. Each offspring contains some chromosomal information from each parent.*

Scientist	Contribution to the Understanding of Inheritance and Heredity
Gregor Mendel	• Analyzed 7 traits with garden peas • Noticed that each individual has two alleles for each trait which segregate during gamete formation • The alleles for one trait assort independently for the alleles for another trait • When two different alleles are present, one appears to be dominant over the other and is always expressed
Walter Sutton	• Helped develop the Chromosome Theory of Heredity; Genes for traits are located on chromosomes
Thomas Hunt Morgan	• Discovered that sometimes alleles for different traits do not assort independently and can be linked because they exist on the same chromosome • Discovered that sex-linked traits appear in different rates in males and in females because males need only one recessive allele to express the recessive trait

- *Many human disorders can be explained by simple genetic inheritance patterns, providing a means to identify future events.*
    - **Recessive Inheritance.** A disease is expressed by an individual who is homozygous recessive (aa).
    - **Dominant Inheritance.** An individual who is homozygous dominant (AA) or heterozygous (Aa) expresses a disease.
    - **Sex-Linked Recessive Inheritance.** A disease is carried on a sex chromosome (usually X) and is inherited as a recessive (a) trait. Afflicted individuals include homozygous recessive females and males with one recessive allele.
    - **Sex-Linked Dominant Inheritance.** A disease is carried on a sex chromosome (usually X) and is inherited as a dominant (A) trait. Afflicted individuals include homozygote dominant or heterozygote females and males with one dominant allele.

49

o Most traits that are sex-linked are due to alleles carried on the X chromosome. If these traits are due to recessive alleles, they occur more frequently in males because males only have one X chromosome and only need one recessive allele to express the recessive trait. Y-linked traits can appear solely in males.

o **Linked traits** are due to genes being located on the same chromosome. With linked traits, offspring will reflect parental phenotypes more frequently.

o With linked traits, recombinant phenotypes arise due to crossing-over during meiosis. The frequency of these recombinant phenotypes may deviate from what would be expected with independent assortment. If the linked genes are far apart on the chromosome, the resulting phenotypes will appear to assort independently. If the linked genes are close, the frequency of recombinant phenotypes will be less than the frequency of parental phenotypes.

o The **recombination frequency** can be used to estimate the distance between genes on a chromosome. The recombination frequency is the number of map units between genes. High recombination frequencies suggest a large distance between the two linked genes.

STRUCTURE AND FUNCTION OF DNA

• *Several experiments built an understanding of DNA as the genetic material. These experiments demonstrate the consistency of the DNA structure and its flexibility and ability to change through inheritance.*

Contribution to Understanding the Function and Structure of DNA	Scientist
• Transformation of genetic material is possible through use of two pneumonia strains	Frederick Griffith
• Repeated Griffith's experiments to find that the transforming agent was DNA	Oswald Avery, Maclyn McCarty, Colin MacLeod
• Used bacteriophages to determine that the DNA from the phage transforms the bacterial host	Alfred Hershey and Martha Chase
• The percentage of the four types of nucleotides is different in different organisms • Percent of adenine nucleotides is the same as the percent of thymine nucleotides • Percent of cytosine nucleotides is the same as percent of guanine nucleotides	Erwin Chargaff
• Operated a lab where X-ray crystallography was conducted to analyze molecular structure	Maurice Wilkins

• Took the first X-ray crystallography picture of DNA showing two repeating patterns.	Rosalind Franklin
• Proposed the double-helix model of DNA based upon contributions of earlier scientists • Proposed semi-conservative model of replication, but lacked supporting evidence	James Watson and Francis Crick
• Used isotopes of nitrogen to demonstrate that DNA replication could not be conservative or dispersive • Evidence that DNA replication is semi-conservative	Matthew Meselson and Franklin Stahl
• Using bread mold were able to discern that each gene appears to be responsible for making one enzyme that is needed for a biological process	George Beadle and Edward Tatum

DNA REPLICATION

- *In the process of DNA replication, a parental DNA double helix is copied precisely and two identical DNA double helices are formed.*
    - In the process of DNA replication, the enzyme **helicase** separates the two strands of the DNA double helix and single-strand binding proteins hold the two strands apart.
    - The **replication fork** is at this point of separation. The DNA strand that runs 3' to 5' is copied directly and the new strand from this template is called the **leading strand**. The DNA strand that runs 5' to 3' is copied in fragments and the new strand from this template is called the **lagging strand**.
    - **DNA polymerase (III)** is the primary enzyme that will build the new DNA strand. It has two major limitations; it cannot begin the replication without a primer and it can add new nucleotides only to a free 3' end of growing DNA strand.
    - On both DNA strands, **RNA primase** adds a small RNA primer to serve as the beginning nucleotide sequence.
    - On the leading strand, DNA polymerase adds nucleotides one at a time to the free 3' end in a direct fashion to the small RNA primer.
    - On the lagging strand, DNA polymerase will add nucleotides to primers that are runnung in the reverse direction. The new DNA strands are built in small fragments so that each one can form from 5' to 3'. These fragments are called **Okazaki fragments**. **DNA ligase** builds covalent bonds between the Okazaki fragments to form one continous strand.
    - After the strands are copied, a different DNA polymerase removes the RNA primer and replaces it with DNA. Then, ligase seals the new piece to the new DNA strand.

o DNA polymerase proofreads the new strands and repairs any mismatch that may have happened.

PROTEIN SYNTHESIS

• ***This two-step process includes transcription and translation. In transcription, one strand of parental DNA is precisely copied into RNA. In translation, the RNA is precisely copied into a polypeptide sequence.***

   o The process of **transcription** occurs in the nucleus of the cell where the DNA template is copied into an RNA transcript. The DNA strand that runs 3' to 5' is the template strand. The other DNA strand is not copied.

   o Small gene sequences are transcribed at one time; not the entire genome.

   o The RNA strand is built much like the leading strand in DNA replication but with RNA nucleotides being added. The process continues until the **terminator** sequence of the template strand is transcribed. At that point, the RNA strand is released and the two DNA strands reanneal.

   o **RNA polymerase** separates the DNA double helix, adds RNA nucleotides to the free 3' side of a growing RNA strand, and reseals the DNA strands back together.

   o The RNA polymerase begins the process when it binds to the promoter sequence of the template strand. **Transcription factors** bind at the TATA box to promote the binding of RNA polymerase to the promoter sequence.

   o The RNA transcript is modified in the nucleus by adding a poly A tail to the 3' side and a modified guanine cap to the 5'side to protect the RNA from degradation in the cytosol and promote translation by ribosomes.

   o **Spliceosomes** cut out introns and fuse the exons together. At this point, the primary transcript is mRNA and can leave the cell nucleus and move to a ribosome for translation.

   o The ribosome has two subunits: 30s and 70s. The two subunits come together when a mRNA strand is present. The two subunits join around the mRNA so that a start codon is docked in the P site of the ribosome.

   o A complete ribosome has three binding sites: the A site (aminoacyl tRNA binding site), the P site (peptidyl tRNA binding site), and E site (exit of empty tRNA).

   o A tRNA with the first amino acid binds to the P site of the ribosome. A tRNA with the next amino acids binds to the A site of the ribosome.

   o The ribosome catalyzes a peptide bond between the two amino acids as the bond between the tRNA and the first amino acid in the P site is broken.

   o The ribosome shifts position so that the tRNA with the growing amino acid chain moves from the A site to the P site. The tRNA that was in the P site is now in the E site. It is released from the ribosome so that it can be rematched to an amino acid.

   o This process continues until a stop codon appears in the A site. At that point, no tRNA binds. Instead, a release factor binds in the A site and it triggers the dissociation of the translation complex.

o The polypeptide that is formed goes through some modifications and folding to form a functional protein

CELL DIVISION

- *DNA replication, and subsequently the process of cell division, is efficient because of cellular packaging.*
  - o **Chromatin** is a DNA-protein complex. Chromatin condenses as the cell prepares for cell division.
  - o There are two forms of chromatin, heterochromatin and euchromatin in eukaryotes.
    - **Heterochromatin** is highly compacted, untranscribed DNA. It is visible during Interphase.
    - **Euchromatin** is a less compacted form of chromatin, usually present during transcription.
  - o Duplicated chromosomes consist of two **sister chromatids**, and when condensed, have a region of narrowing called the **centromere**.
  - o The process of cell division pulls sister chromatids apart and provides each cell with a complete chromosome set.
    - Every species has a specific number of chromosomes.
    - Human **somatic cells** (body cells) have 46 chromosomes. **Gametes** (reproductive cells) have 23 chromosomes.

- *Continuous cell division involves the distribution of identical DNA to two daughter cells.*
  - o The cell cycle consists of two main phases: mitosis (which includes cytokinesis) and interphase.
  - o Mitosis and interphase are further divided into subphases.
  - o Interphase has the longest duration in the cell cycle and has three subphases: G1, S, and G2.
    - Interphase G1: Cell growth and preparation
    - Interphase S: DNA Synthesis
    - Interphase G2: Final preparations for cell division. Asters are formed. Chromosomes are in the form of chromatin.
    - Prophase of Mitosis: Chromosomes (as sister chromatids) are highly condensed, visible and the nucleoli disappear. Spindle begins to form.
    - Prometaphase of Mitosis: Nuclear envelope breaks apart. Chromatids have a kinetochore, located at the centromere region to attach to the microtubules.
    - Metaphase of Mitosis: Chromosomes are present on the metaphase plate. Kinetochores attach to the microtubules.
    - Anaphase of Mitosis: Centromeres of each chromosome separate. Kinetochore microbtubules shorten, and nonkinetochore microtubules lengthen.
    - Telophase of Mitosis: Nuclear envelope begins to reform. Chromatin becomes less tightly wound. Cytokinesis usually begins during telophase.
    - Cytokinesis: In animal cells, cytokinesis involves the formation of a

cleavage furrow, pinching the cell into two distinct cells. In animal cells, vesicles from the Golgi bodies of the cell join at the center of the cell, creating a **cell plate**. The vesicles carry cell wall components and two distinct cells are formed.

- *Interphase and mitosis alternate in order to maintain continuity in cell division.*
  - Not all cells are dividing all of the time. Some cells (other than in recent experimental settings) do not divide at all.
  - Growth, development, repair and replacement are primary purposes of cell division.

- *The cell cycle is continuous and is constantly being regulated by a molecular control system.*
  - There are checkpoints in the cell cycle that determines whether or not cell division will occur. If a cell has not completed specific tasks, it will not proceed with cell division. There are known checkpoints at G_1, G_2 and M of interphase.
  - If a cell does not meet the requirements at the G_1 checkpoint, it will move into a state called G_0, a nondividing state.
  - Proteins, called **cyclin dependent kinases**, regulate the cell cycle control system.
  - Kinases are always present in a cell, but in order to be active, they must be bonded to a cyclin. These Cdks fluctuate in concentration throughout the cell cycle.
  - The resulting complex, MPF, causes a passage of the checkpoint in the cell cycle at G_2.

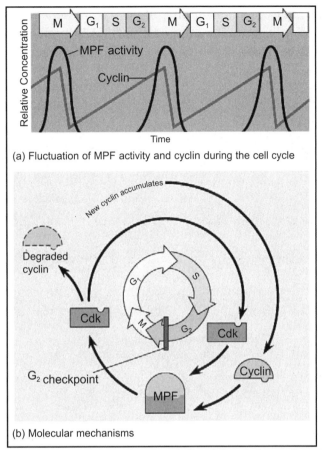

Fig. 12.14, p. 227 from BIOLOGY, 6th ed. by Neil A. Campbell and Jane B. Reece. Copyright © 2002 by Pearson Education, Inc. Reprinted by permission.

- ***The cell cycle control system can malfunction and alter the continuity of cell division.***
    - o Some cells, such as cancer cells, do not respond to control mechanisms. These cancer cells are not density dependent (grow past a complete single layer), perhaps divide independently without growth factors, or may have an abnormal cell cycle control system. Cancer cells do not stop dividing at checkpoints and can often divide indefinitely.

MEIOSIS

- ***Asexual reproduction provides a means for identical genetic material to be passed on to offspring.***
    - o In **asexual reproduction**, one parent passes all of its genetic information to its offspring. The offspring contains a genome that is identical to the parent.
    - o Asexual reproduction does not provide an opportunity for genetic variation, with a recombination of genes.
- ***Sexual reproduction results in offspring that do not have identical complements of DNA.***
    - o In **sexual reproduction**, gametes (sex cells) containing 23 chromosomes (a haploid complement of humans) combine. Sperm cells and ova are gametes.

• *Sexual life cycles involve an alternation of fertilization and meiosis.*

  o Meiosis and fertilization alternate in all sexually reproducing organisms. There
    are three different types of sexual life cycles, all of which allow for genetic
    variability among offspring.

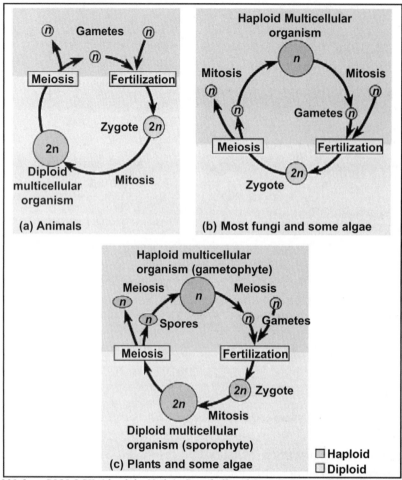

Fig. 13.5, p. 238 from BIOLOGY, 6th ed. by Neil A. Campbell and Jane B. Reece. Copyright © 2002 by Pearson Education, Inc. Reprinted by permission.

• Genetic variation in offspring results from crossing over, random fertilization and
  independent assortment of chromosomes.

  o **Crossing over:** In prophase of meiosis I, homologous chromosomes come
    together to form a **synaptonemal complex**. Crossing over occurs when
    homologous portions of two nonsister chromatids exchange genetic information.
    Therefore, each chromosome of each gamete can contain DNA from two
    different parents.

  o **Random fertilization:** Fertilization is a random process. A single ovum and a
    single sperm, from millions of possibilities, combine to form a zygote with
    trillions of possible combinations of genes.

○ **Independent assortment of chromosomes:** There is a 1 in 2 chance that each gamete will get a maternal chromosomes and the same chance that it will get a paternal chromosome. As this is the case for all 23 chromosomes, the number of possible combinations in the resulting gametes is about eight million.

- *Heritable genetic variations in offspring, which are favored by the environment, make evolution possible.*

- *The segregation of alleles during meiosis provides for continuity of distribution, as well as variation of offspring.*
    ○ Punnett crosses can be used to display this segregation.
    ○ Alleles may be **dominant** or **recessive** (expressed or not expressed).
    ○ **Homozygous** individuals have two of the same alleles for a specific trait (AA or aa). These individuals would express either the dominant (AA) trait or the recessive (aa) trait.
    ○ **Heterozygous** individuals have one of each allele (Aa). If the trait was completely dominant, this individual would express the dominant (A) trait and mask the recessive (a) trait.
    ○ **Phenotype** describes the outward expression of the genotype or what an organism looks like. In complete dominance, homozygous dominant individuals (AA) and heterozygotes (Aa) express the dominant phenotype while homozygous recessive individuals (aa) express the recessive trait.
    ○ **Genotype** describes what alleles an organism carries and is more specific than phenotype.

- *Laws of probability explain the predictable results of simple genetic crosses.*
    ○ Rule of addition: Genetic events can occur independently from one another. The percentage for each event can be added together to determine the sum of probabilities of both events happening.
        ■ Example: What is the chance that Tom and Susan will have a boy OR a girl?

        ■ There is a 50% chance that Tom and Susan will have a boy, and a 50% chance that Tom and Susan will have a girl. Therefore, the chance that Tom and Susan will have a boy OR a girl is 50% + 50%, or 100%.

    ○ Rule of Multiplication: Genetic events occur simultaneously, or not independently from one another. Therefore, the percentage for each event can be multiplied to determine the probability of both events happening.
        ■ Example: What is the chance that Tom and Susan will have a boy as their first child, AND a girl as their second child?

        • There is a 50% chance that Tom and Susan will have a boy as their first child. There is also a 50% chance that Tom and Susan will have a girl as their second child. In order for BOTH of these events to happen, we

multiply the percentage of the 1st event (50%) x the percentage of the 2nd
event (50%) and determine that there is a 25% chance of both events
occurring.

- *Complex gene interactions account for differences from the expected results of simple Mendelian patterns.*
  - Multiple Alleles: Many times there are more than just two forms of a gene.
    Blood type is a common example of this inheritance, as there are three alleles for
    this gene: I^A, I^B, and i.
    - I^A and I^B are codominant, as both are expressed.
    - I^A and I^B are dominant to i.
    - There are three possible alleles, but an individual only carries two alleles.

  - Often times, alleles are not dominant totally dominant over another allele. This
    incomplete dominance results in intermediate phenotypes for the heterozygote.
    - Example: Color in some flowers is not inherited as a completely dominant
      trait. If a homozygous red flower (C^rC^r) is crossed with a homozygote
      white flower (C^wC^w), the resulting offspring will be heterozygotes (C^rC^w).
      They will express a pink phenotype, or an intermediate between the two
      homozygotes.
    - Genes often have the ability to have many phenotypic effects. One gene can
      influence several different characteristics. This phenomenon is called
      pleiotropy.

  - Sometimes one gene influences the expression of another gene. This is called
    epistasis and is a complex gene interaction.
    - Example: In Labrador retrievers, the inheritance of fur coloration is an
      epistatic interaction.
    - One gene specifies coat color in typical dominant fashion. The dominant
      allele (B) causes an expression of black coat color, and the inheritance of the
      recessive allele (b) results in brown coat color.
    - However, a second gene must be present in order for the pigment that causes
      coat color to be produced. Without a specific enzyme (the second gene),
      melanin will not be deposited, and regardless of the allele that is inherited,
      the dog will not be black or brown, it will be a yellow Labrador.

  - **Polygenic inheritance** is a type of inheritance where many genes function
    together to form a continuum within a population for a specific trait.
    - The additive effects of the genes determine a particular phenotype.
    - Example: skin color. Individuals who carry the following genes would have
      dark skin (AABBCC) whereas heterozygous individuals (AaBbCc) are not as
      dark and homozygous recessive individuals (aabbcc) have extremely fair
      skin.
    - The number of dominant or recessive alleles is what is important in this
      system. For example, AABbcc has the same effect as AaBbCc, as each has
      three dominant alleles and three recessive alleles.

MULTIPLE CHOICE QUESTIONS

1. What is the probability that Mark's and Cindy's first-daughter will be color-blind if Cindy is heterozygous for this sex-liked condition and Mark has normal vision?

   (A)   0 %
   (B)   25 %
   (C)   50 %
   (D)   75 %
   (E)   100 %

2. In the F2 generation, a 9:3:3:1 phenotype ratio was predicted. However, most of the resulting offspring were dominant for both traits or recessive for both traits. Few offspring were dominant for one trait and recessive for the other. This result would best be explained by

   (A)   sex linkage
   (B)   codominance
   (C)   linkage
   (D)   epistasis
   (E)   pleiotropy

3. Griffith's experiment with pneumonia and mice demonstrated that

   (A)   transformation of the genetic material is possible.
   (B)   DNA was the genetic material.
   (C)   protein was the genetic material.
   (D)   genes are located on chromosomes.
   (E)   traits for smooth and rough coatings on pneumonia-causing bacteria are linked.

4. In Chargaff's findings, the percentage of adenine nucleotides in an organism was equal to the percentage of

   (A)   guanine nucleotides
   (B)   thymine nucleotides
   (C)   cytosine nucleotides
   (D)   uracil nucleotides
   (E)   adenine nucleotides in all other living organsms

5. During DNA replication, the enzyme that separates the DNA double helix is

   (A) helicase
   (B) DNA ligase
   (C) DNA polymerase
   (D) RNA polymerase
   (E) RNA primase

6. All of the following enzymes are involved in DNA replication EXCEPT for

   (A) helicase
   (B) DNA ligase
   (C) DNA polymerase
   (D) RNA polymerase
   (E) RNA primase

7. In the transcription process, which strand(s) of the parental DNA act(s) as templates for RNA production?

   (A) both strands of the DNA double helix
   (B) the strand that runs from 5' to 3'
   (C) the strand that runs from 3' to 5'
   (D) either one strand is selected randomly by the polymerase enzyme
   (E) the strand that has a larger number of A nucleotides

8. All of the following molecules are involved in the translation process EXCEPT for

   (A) DNA
   (B) mRNA
   (C) tRNA
   (D) rRNA
   (E) amino acids

9. The genetic code

   (A) varies greatly from one organism to the next
   (B) is highly conserved within a large group, but has changed dramatically between large groups; i.e. conserved in animals different between plants and animals
   (C) is highly conserved in all organisms
   (D) outlines how to convert DNA into RNA
   (E) outlines how to convert mRNA into DNA

10. Proteins are translated at a

    (A)   golgi complex
    (B)   lysosome
    (C)   vacuole
    (D)   nucleolus
    (E)   ribosome

11. Sickle cell anemia is an inherited disease caused by a recessive allele. If there is 50% chance for the offspring to have the disease, and the mother is a carrier, what is the genotype of the father?

    (A)   Aa
    (B)   aa
    (C)   AA
    (D)   Aa or aa
    (E)   AA or Aa

12. A mother has type O blood and a father is heterozygous for type B blood. What are the chances of the offspring having type O blood?

    (A)   0%
    (B)   25%
    (C)   50%
    (D)   100%
    (E)   unable to determine the offspring

13. In the following pedigree, darkened individuals express a particular genetic disease. What is the probable mode of inheritance in the following pedigree?

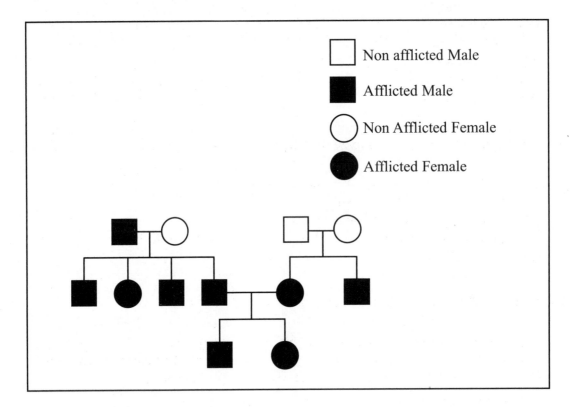

(A)   sex-linked dominant
(B)   sex-linked recessive
(C)   autosomal recessive
(D)   autosomal dominant
(E)   unable to determine the mode of inheritance

14. DNA synthesis occurs during which phase of the cell cycle?

(A)   G1
(B)   S
(C)   G2
(D)   M
(E)   cytokinesis

15. Condensation of chromatin fibers into heterochromatin occurs during which of the following phases?

   (A)   Interphase
   (B)   Metaphase
   (C)   Anaphase
   (D)   Prophase
   (E)   Telophase

16. What causes the MPF complex to initiate mitosis in the cell cycle?

   (A)   accumulation of cyclins associated with CDKs
   (B)   CDKs reacting with the MPF
   (C)   decrease in the cyclin concentration
   (D)   production of enzymes to catalyze adequate amounts of MPF
   (E)   increase in kinases

17. Which of the following statements is correct?

   (A)   DNA synthesis occurs during the S phase of interphase.
   (B)   Mitosis and meiosis result in daughter cells that are identical to parent cells.
   (C)   DNA synthesis occurs once in mitosis and twice in meiosis.
   (D)   Crossing over occurs in anaphase of meiosis.
   (E)   Meiosis results in diploid daughter cells.

18. When light spored Sordaria are crossed with dark spored Sordaria, the following results are obtained. From these results, how many of the arrangements resulted from crossing over?

            3 with 2 light/4 dark/2 light
            5 with 2 light/2 dark/2 light/2 dark
            4 with 2 dark/2 light/2 dark/2 light
            8 with 4 light/4 dark
            9 with 4 dark/4 light

   (A)   9
   (B)   12
   (C)   17
   (D)   29
   (E)   8

19. Where does the checkpoint regulated by MPF occur in the cell cycle?

    (A)   after G1
    (B)   after metaphase
    (C)   after S
    (D)   after G2
    (E)   after telophase

20. Four genes, A, B, C and D, have loci on the same chromosome. Using the following crossover frequencies, make a map of the chromosome.

A-D	10%
B-C	20%
A-C	15%
C-D	25%
B-D	45%

    (A)   ABCD
    (B)   BCAD
    (C)   ADCB
    (D)   DABC
    (E)   ABDC

FREE RESPONSE QUESTIONS

1. The following data was collected using *Drosophila melanogaster* and two traits. In the parental cross, flies that were homozygous dominant for both traits were crossed with flies that were homozygous recessive for both traits. The two traits that were analyzed were wild type wings (dominant) vs. vestigial wings (recessive) and red eyes (dominant) vs. cinnabar eyes (recessive). After the initial cross, F_1 flies were crossed.

Phenotypes	Parent Flies	F_2 Flies	F_2 Flies
Wild type wings, red eyes	500	1200	985
Wild type wings, cinnabar eyes	0	0	45
Vestigial wings, red eyes	0	0	35
Vestigial wings, cinnabar eyes	500	0	310
Totals	1000	1200	1400

   a.   What is the most probable explanation for how these two traits are inherited?
   b.   Why does this data suggest that mode of inheritance?
   c.   Is there evidence of recombination between these two traits? **Explain**.
   d.   **Explain** how this mode of inheritance differs from the conclusions of Gregor Mendel.

2. Mitosis and meiosis are both integral processes in gene transmission. However, the processes, while similar, have distinct differences. For these processes:

   a.   **Explain** how alleles are transmitted differently in mitosis vs. meiosis and why both processes are necessary.
   b.   Genetic variation in offspring is a result of several processes. **Explain** these processes at the chromosomal level.
   c.   **Discuss** how meiosis impacts the evolution of species.

NO TESTING MATERIAL PRINTED ON THIS PAGE

GO ON TO THE NEXT PAGE →

Theme Five

Relationship of Structure to Function

THE CHEMICAL CONTEXT OF LIFE

- *The behavior of an element is determined by its atomic structure.*
  - The atom is composed of three subatomic particles: **neutrons, protons, and electrons**. Electrons are negatively charged, protons are positively charged and neutrons are electrically neutral.
  - The fact that electrons and protons are oppositely charged is what maintains the stability of the atom, keeping the electrons near the nucleus.
  - The number of protons in an atom is the **atomic number** of the element.
  - The **mass number** is the number of protons and neutrons in the atom.
  - The mass of an atom is called the **atomic weight**.

- *A difference in the number of neutrons of an atom accounts for different behaviors.*
  - **Isotopes** are forms of an element that have the same number of protons but a different number of neutrons. This difference may cause the nucleus to spontaneously decay, giving off particles, and eventually decaying to form another element, with the loss of protons.
  - Biologically, these **radioactive isotopes** can be used in metabolic tracing and fossil dating.

- *The chemical function of an atom is clearly determined by its electron configuration.*
  - Electrons are found in **electron shells** or **energy levels**. The most important electron shell in discussing the behavior of an atom is the outermost shell, where the electrons are termed **valence electrons**.
  - These valence electrons determine the bonding patterns for each atom, by either sharing or transferring electrons to form a chemical bond.

- *The bonding patterns of atoms determine their strength and function.*
  - The different types of chemical bonds include: **covalent bonds**, where electrons are shared between atoms, and **ionic bonds**, where electrons are transferred between atoms, creating **ions**.
    - Covalent Bonds: biologically important in forming organic molecules and high energy molecules.
    - Ionic Bonds: biologically important in forming salts. The strength of these bonds is largely influenced by the environment. Ionic bonds are much stronger in dryer environments as compared to when they are placed in water.

■ Hydrogen Bonds: biologically important in water molecules and in dissolving particular molecules.
■ Van der Waals Interactions: biologically important in proteins, causing temporary charges that allow molecules to stick to one another.

- *The biological function of a molecule is directly related to its shape.*
  o The shape of a molecule is critical. In biological molecules, form fits function and is integral in explaining how it is recognized by other molecules.
  o Changing the shape of a molecule generally renders the molecule useless in biological recognition.
    ■ Receptors on the outside of cells are complimentary to the shape of particular molecules.
    ■ If this shape changes, due to a change in bonding patterns, the molecule will not be recognized by the receptor, and a particular biological process will not take place.

WATER

- *Water is a biologically significant molecule due to its polarity.*
  o **Polar molecules** are molecules where the ends of molecules have opposite partial (positive and negative) charges, causing the molecules to engage in **hydrogen bonding**.

- *Water functions as a temperature stabilizer.*
  o Water has a **high specific heat**. It takes a great deal of energy to raise the temperature of water.
  o Since the specific heat of water is so high, large amounts of energy can be absorbed or released without a large change in temperature.
  o Because of strong intermolecular forces, including hydrogen bonding, the **heat of vaporization** of water is very high. As the particles with higher energy leave the system, they carry the heat energy with them, therefore evaporation of water from the surface of an organism cools the organism

- *Water molecules behave as a highly structured liquid due to the fact that hydrogen bonds between water molecules break and reform frequently.*
  o There is a constant breaking and reforming of the hydrogen bonds between water molecules.
  o Therefore water molecules are nearly always bonded to one neighboring water molecule or another. This is called **cohesion**.
  o **Cohesion** is biologically important in transporting water through the vessels of plants.
    ■ When transpiration occurs at the surface of a leaf, and a water molecule is evaporated, the cohesiveness of water will pull another molecule upward in the vessel while it is attached to an additional molecule. This process will continue to the roots, engaging all water molecules in hydrogen bonding and overcoming the force of gravity as water moves from the roots to the leaves.

o **Adhesion**, or the attraction of water molecules to the vessel walls, also helps water to overcome gravity.
o Hydrogen bonding also gives water the property of **surface tension**, which comes from the ordered arrangement of water molecules hydrogen bonded together at the surface. The breaking of the surface of water is resisted due to the constant breaking and reforming of hydrogen bonds. This allows some animals to walk across water without breaking the surface.

• *When water freezes, its crystalline lattice shape causes it to be less dense than as a liquid.*
o As water molecules freeze, the hydrogen bonds keep the molecules far away from one another, decreasing the density of water. This allows ice to float on water.
o Floating ice insulates the water underneath it, creating a hospitable environment for the aquatic life below.
o Biologically this is significant because if ice were more dense than water, aquatic life would cease to exist, as the entire ecosystem would freeze solid.

• *Hydrogen bonding results in water being a very versatile solvent.*
o The positively charged hydrogen regions of water are attracted to the negatively charged region of **solute** molecules, or molecules that are being dissolved. The negatively charged oxygen regions of water are attracted to the positively charged regions of solute molecules, forming a **hydration shell**.
o Any molecule that is ionic or polar can be dissolved in water. These **hydrophilic** molecules are said to be water loving.
o Other molecules are not attracted to water. These molecules are considered to be **hydrophobic** and are generally nonionic and nonpolar molecules.

CARBON AND ORGANIC MOLECULES

• *Carbon is the central atom in organic molecules. Because of carbon's ability to form four covalent bonds in a variety of shapes and patterns, it adds strength and support to large organic macromolecules.*
o There are common functional groups attached to the carbon skeleton of organic molecules that leads to the basic properties of the overall molecule.

Functional Group	Structure	Properties and Compounds
Hydroxyl	--OH An oxygen bonded to a hydrogen	• Compounds are called alcohols • Is polar due to electronegative oxygen bound to hydrogen
Carbonyl	--CO A carbon double bonded to an oxygen	• When group is in middle of carbon chain, the compound is a ketone. • When at the end, compound is an aldehyde
Carboxyl	--COOH A carbon double bonded to an oxygen and bound to a hydroxyl	• Organic acids contain this group—called carboxylic acids • Hydrogen ions tend to dissociate because of polarity giving acidic properties
Amino	--NH$_2$ A nitrogen bonded to two hydrogens	• Compounds with an amino group are called amines. • Acts as a base
Sulfhydryl	--SH A sulfur bonded to a hydrogen	• Compounds with a sulfhydryl are called thiols. • Involved in maintaining protein structure
Phosphate	--PO$_4$ A phosphorous bonded to four oxygens	• Is an anion • Involved in nucleotides, ATP, and phospholipids

- *Carbon is the central component of all four major organic macromolecules.*
  - **Carbohydrates** are composed of monomer subunits called monosaccharides. Monosaccharides are simple sugars like glucose, fructose, and ribose. These compounds provide energy.
  - Two monosaccharides link together through dehydration synthesis to form a disaccharide. The resulting covalent bond is called a **glycosidic linkage**.
  - Examples of disaccharides include sucrose and maltose.
  - Many monosaccharides linked together form polysaccharides like cellulose, starch, chitin, and glycogen. These compounds add structural support and provide storage.
  - **Lipids** are nonpolar, hydrophobic compounds. Many lipids are composed of fatty acid monomers. All lipids consist primarily of carbon and hydrogen resulting in the nonpolar structure.
  - A fatty acid has a long hydrocarbon chain with a carboxyl group at the end.
  - Fatty acids with multiple bonds between carbons in the hydrocarbon are **unstaturated** and fatty acids without these multiple bonds are **saturated**.
  - A fat contains three fatty acids bound to a three-carbon glycerol. The fatty acids join to the glycerol through dehydration synthesis. The resulting covalent bond is an ester linkage. Fats are storage compounds.

o A phospholipid contains a glycerol with two fatty acids and a phophate group. Phospholipids have a polar phosphate group and two nonpolar fatty acid tails. This amphipathic nature of a phospholipids make it an integral component of cell membranes.

o A steroid contains four fused rings of carbon and hydrogen. These compounds often act as hormones or chemical regulators. Examples include cholesterol and testosterone.

o **Nucleic acids** are composed of subunits called **nucleotides**. There are two common nucleic acids: DNA and RNA.

o In DNA, there is a five-carbon sugar (deoxyribose) that is bound to both a phosphate group and a nitrogenous base (adenine, cytosine, thymine, or guanine).

o In RNA, there is a five-carbon sugar (ribose) that is bound to both a phophate group and a nitrogenous base (adenine, cytosine, uracil, or guanine).

o DNA has two **antiparallel** strands that are connected through hydrogen bonding. The adenines of one strand bond to the thymines of the other strand through two hydrogen bonds. The cytosines of one strand bond to the guanines of the other strand through three hydrogen bonds.

o Cytosine, thymine, and uracil are pyrimidines (they have one six-member ring of carbon and nitrogen).

o Adenine and guanine are purines (they have a six-member ring of carbon and nitrogen fused to a five-member ring of carbon and nitrogen).

o **Proteins** have the most diverse shapes and functions. They are composed of an intricate arrangement of amino acids. The complex three-dimensional shape of a protein determines its function.

o There are twenty different types of amino acids that are strung together to make any protein. An amino acid has a central carbon bound to a carboxyl group, an amino group, a hydrogen, and an R group. The R group is the variable component. It is what distinguishes the twenty types of amino acids.

o Two amino acids fuse through **dehydration synthesis**. The resulting covalent bond is called a **peptide bond**.

o The peptide strand that forms goes through an intricate folding process, often assisted by chaperone proteins to yield a functional protein.

o The **primary structure** of a protein is the sequence of amino acids in the protein.

o The **secondary structure** of a protein is the regular interactions between amino acids that cause the formation of alpha helices and pleated sheets.

o The **tertiary structure** of a protein is the subsequent folding that follows this regular secondary folding. It is reinforced by intricate interactions between amino acid R groups.

o The **quaternary structure** of a protein is only found in proteins that contain more than one subunit. These subunits come together through a folding process that yields the quaternary structure.

CELL STRUCTURE

- *The composition of the cell membrane gives it the unique property of being a selective barrier for cells.*
  - o The cell membrane is composed primarily of a **phospholipid bilayer**. This bilayer is composed of molecules that are both **hydrophilic** and **hydrophobic**. This forms a stable boundary between two environments that are largely composed of water, as the hydrophilic portions are near water, separated by a hydrophobic portion.

- *Proteins and carbohydrates within the cell membrane are important in permeability and recognition.*
  - o **Integral proteins** span the entire length of the phospholipid bilayer and are important in passive and active transport processes.
  - o **Peripheral proteins** are bound to the outside of the bilayer providing strength to the cell.
  - o Membrane carbohydrates are integral in cell to cell recognition. These "name tags" allow cells to recognize one another, prevent rejection, and enhance sorting.

- *Many organelles (excluding ribosomes) within a cell are surrounded by a membrane allowing specific functions to take place within each organelle.*
  - o These membranes provide for specialized environments within each organelle. Therefore processes that may otherwise not be able to occur at the same time within the same cell, can occur in specific organelles.

- *The nucleus of a eukaryotic cell is surrounded by a double membrane, perforated with pores, allowing passage of materials into and out of the nucleus.*
  - o The **nuclear lamina** maintains the shape of the nucleus.
  - o The nucleus contains **chromatin** (DNA – protein complex), chromosomes (discernable DNA organized into structures), and a nucleolus (site for the synthesis of rRNA).

- *The location of ribosomes within the cell greatly impacts the type of proteins they produce.*
  - o Ribosomes are the sites of protein synthesis within a cell. They may be either **bound** or **free** ribosomes
    - ▪ Bound ribosomes produce proteins that are for export or for membrane support. These ribosomes are located on the endoplasmic reticulum
    - ▪ Free ribosomes produce proteins that will be used within the cell. These ribosomes are floating freely within the cell.

- *The structure of the endoplasmic reticulum directly relates to its function.*
  - If the endoplasmic reticulum is covered with ribosomes, it is considered to be **rough endoplasmic reticulum**. This organelle functions in the synthesis of secretory proteins.
  - The membrane of the endoplasmic reticulum surrounds these proteins as they travel in vesicles to the Golgi apparatus.
  - If the endoplasmic reticulum is not covered with ribosomes, it is considered to be **smooth endoplasmic reticulum**. This organelle functions in many metabolic processes.
    - Detoxification of drugs
    - Synthesis of hormones
    - Synthesis of lipids for new membranes
    - Carbohydrate metabolism

- *The distinct polarity of the Golgi apparatus allows shipping and receiving to take place at one organelle.*
  - The **Golgi apparatus** functions as a packaging center within the cell.
  - The *cis* face of the Golgi faces the endoplasmic reticulum and receives vesicles from the ER.
  - As the products from the ER travel from the *cis* to the *trans* face, they are modified. This may include adding molecular identification tags, modifying glycoproteins, or other modifications.
  - The products are secreted in vesicles off of the *trans* face of the Golgi apparatus to travel to other sites.

- *The membrane of the lysosome allows for macromolecule digestion without destruction of the cell.*
  - The cytoplasm of the cell has a relatively neutral pH. **Enzymes** required for macromolecule digestion have an optimum pH of approximately 5. The lysosomal barrier allows for these enzymes to function without disrupting the cytoplasmic pH of the cell and disrupting other metabolic processes and destruction.

- *Mitochondria maintain a very high surface area that increases the efficiency of the organelle.*
  - The **mitochondria** are the site of cellular respiration in the cell. ATP is produced from the breakdown of other fuels, such as sugars and fats when oxygen is present.
  - The inner membrane of the mitochondria is highly folded, with folds being called **cristae**. These cristae increase the surface area of the mitochondria.
  - By increasing the surface area of the mitochondria, cellular respiration can occur at a much higher rate, increasing the efficiency of the cell.

- *The compartmentalization of the chloroplast increases the efficiency of photosynthesis within the cell.*
    - o The chloroplast is divided into a **thylakoid space** and the **stroma**.
    - o **Chlorophyll**, a pigment, is embedded in the thylakoid membrane of the chloroplast. When these pigments are struck by photons of light, energy is passed from molecule to molecule until it reaches a reaction center.

- *The diverse structure of the cytoskeleton functions to provide support, motility and regulation to the cell.*
    - o The main components of the cytoskeleton are **microtubules, microfilaments, and intermediate filaments.**
    - o Each component, due to its different structure provides a different function to the cell.
    - o **Microtubules** are hollow tubes, relatively large in diameter. They are composed of the protein tubulin and function primarily in maintaining cell shape, cilia, flagella, and cell division.
        - ▪ **Cilia** and **flagella** are structures that give motor properties to some cells.
        - ▪ Cilia usually occur in large numbers on the cell. They are short in length and work to move fluid over the surface of a cell.
        - ▪ Flagella usually occur singly, or possibly as just a few per cell. They function to propel cells (or unicellular organisms) through water.
    - o Microfilaments are composed of actin and are the smallest of the components of the cytoskeleton. They are composed of the protein actin and function primarily in maintaining cell shape, muscle contraction, cytoplasmic streaming and the formation of cleavage furrows during cell division.
    - o Intermediate filaments are fibrous proteins of intermediate size. They are usually composed of keratin and function to maintain cell shape as well as anchor the nucleus and form the nuclear lamina.

GENETICS OF VIRUSES AND BACTERIA

- *As small, noncellular organisms, viruses owe many of their properties and dimensions to the simple genetic material that they contain.*
    - o The structure of the viral genome allows host penetration and often the appearance of host disease.
    - o Viruses have a protein coat with an internal genetic material.
    - o The genetic material can be double-stranded DNA (adenovirus, papovavirus, herpesvirus, and poxvirus), single-stranded DNA (parvovirus), double-stranded RNA (reovirus), single-stranded RNA that acts as mRNA (picornavirus, coronavirus, flavivirus, and togavirus), single-stranded RNA that serves as a template for mRNA synthesis (filovirus, orthomyxovirus, paramyxovirus, and rhabdovirus), or single-stranded RNA that acts as a template for DNA (retrovirus).

- o Viruses have two common life cycles: **lytic cycle** and **lysogenic cycle**.
- o In a lytic virus, the virus injects its genetic material into the host cell. The genetic material transforms the host cell into a factory that produces viral genetic material and proteins. The genetic material is packaged inside the proteins and new viruses are synthesized. Then, the host cell is lysed and destroyed and the new viruses emerge to find new hosts.
- o In a lysogenic virus, the virus injects its genetic material into the host cell. The viral genetic material gets incorporated into the host genome where it is copied into all host daughter cells. Then, when the environment changes, the virus leaves the lysogenic cycle and transforms the host cells. The host cell synthesizes new viruses that cause the destruction of the host cell. New viruses are released to find new host cells.
- o **Viroids** are common plant blights that consist of infectious pieces of RNA; there is no protein component to its structure.
- o **Prions** can cause animal infections. These infectious proteins model functional proteins and often result in neurological damage. The structure is simply a modified protein; it does not have any genetic material.

- • *Bacteria have simple genomes that allow for some levels of variability even though there is no sexual reproduction.*
  - o Bacteria have one circular chromosome. It is small and consists of almost all coding DNA.
  - o Often bacteria have small circular pieces of extragenomic DNA called **plasmids**. These plasmids contain active genes that can benefit the bacteria.
  - o Bacteria reproduction is through **binary fision**. There is no genetic exchange or recombination during reproduction.
  - o Variation can happen to a bacterial genome through mutation, conjugation, transformation, or transduction.
  - o **Mutation** allows for altered alleles. As asexual organisms, the mutations appear in all daughter cells and are much more of a direct source of variation than in sexually-reproducing organisms.
  - o **Transformation** involves the uptake of foreign DNA from the environment. The DNA binds to surface proteins before it is pulled into the cell and incorporated into the cell's genome or acts as a plasmid.
  - o **Conjugation** involves one plasmid-containing bacterium copying that plasmid and transferring a copy to another bacterium. In the process, the donor cell forms a cytoplasmic bridge called a pilus where the plasmid moves across to the recipient cell.
  - o In **transduction**, a virus carries a piece of bacterial DNA from a previous host to its new host.
  - o Bacteria regulate genetic activity through **operons**. An operon is a simple regulatory system that uses an operator and an associated repressor protein.
  - o When a **repressor** is bound to the operator, it prevents RNA polymerase from reaching the promoter and the gene is inactive.
  - o If an **inducer** does not allow the bonding of the repressor to the operator, the promoter is available to the RNA polymerase and the gene is active.

PLANT STRUCTURE AND NUTRITION

- *The diverse structure of leaves allows for many different functions within the Plant Kingdom.*
    - o Some leaves are extremely large or doubly compound to withstand tearing and provide barriers to specific pathogens. If a pathogen only invades one leaflet, rather than an entire leaf, the plant may have a lesser chance of destruction.
    - o **Tendrils** are used to climb or cling to supports by some plants.
    - o **Spines** allow for photosynthesis to take place without extensive water loss.
    - o **Succulent plants** have leaves that are very thick and can store large amounts of water.
    - o Leaves may be very bright in color, attracting pollinators.

- *The structure of the root system of a plant affects its ability to survive in specific environments.*
    - o **Fibrous root systems** extend below the soil surface. This provides a great anchor for the plant and allows the plant to obtain a great amount of water that is close to the surface.
    - o **Tap root systems** consist of one large root, primarily for anchoring purposes. These taproots often store reserve food and allow plants to access water well below the surface of the soil.

PLANT TRANSPORT (LAB #9)

- *The structure of guard cells plays a role in balancing the processes of photosynthesis and transpiration.*
    - o **Stomata** are pores in the epidermis of a leaf surrounded by a pair of **guard cells**. These guard cells adjust the size of the stomata, thus regulating water loss in a plant.
    - o When guard cells take in water they swell. As the cell accumulates K+ ions, the water potential of the guard cell drops, and water flows inward. This forces the stomatal opening to become larger, allowing a release of water.
    - o When there is a lack of K+ in the guard cells, and water conservation is at a premium, the guard cells are flaccid, and the stomata remain closed, preventing the escape of water.
    - o Often times, light stimulates the opening of the stomata, induced by a swelling of the guard cells.
    - o When there is a lack of carbon dioxide in the leaf, the stomata will also open.
    - o **Circadian rhythms** also affect the opening and closing of stomata, keeping stomata open during the day and closed at night.
    - o Water deficiency of a plant will also cause guard cells to lose their turgor pressure, closing stomata. This slows photosynthesis, but prevents death of the plant.

- High temperatures and excessive transpiration may also cause the guard cells to become flaccid, thus ceasing photosynthesis in the effort to balance water loss.

- *The structure of aquaporins facilitates water transport rates a cross membranes.*
  - **Aquaporins** are protein channels that open and close in reponse to the water content of a plant cell. These aquaporins do not necessarily affect the water potential gradient, but the rate of water uptake or loss.

- *The compartments of plant cells provide for the maintenance of cell shape.*
  - The **cell wall, cytosol** and the **vacuole** are the three main compartments of most plant cells.
  - The **tonoplast** surrounds the vacuole and monitors the passage of molecules into and out of the cytosol.

- *The presence of root hairs and mycorrhizae increase the efficiency of water absorption.*
  - **Root hairs** greatly increase the surface area of the root.
  - A soil solution passes into the epidermal cells of the root, into the cortex, where it meets the **Casparian strip** and flows through the selectively permeable **symplast**.
  - Partnerships between roots and symbiotic fungi are called **mycorrhizae**, and work to absorb water and minerals due to a greatly increased surface area. This dramatically increases the efficiency of water absorption of the plant.

- *The selective permeability of the endodermis regulates absorption of water.*
  - The **Casparian strip**, a waxy strip regulates the absorption of water. The only way that dissolved minerals can cross this barrier is to go through the **symplast**, which contains a selectively permeable membrane.
  - This allows the roots to be selective in which minerals they absorb and which minerals they do not.

PLANT REPRODUCTION

- *All plants display an alternating life cycle. Each plant possesses a multicellular, haploid gametophyte stage and a mulitcellular, diploid sporophyte stage.*
  - **Gametophyte** tissue is haploid and multicellular. It is the dominant tissue type in bryophyte plants.
  - **Sporophyte** tissue is diploid and multicellular. It is the dominant tissue type in all vascular plants.
  - In a typical plant life cycle, the haploid sperm and the haploid egg fuse to make a diploid zygote. This zygote undergoes mitosis to become a multicellular embryo which continues mitosis to become a multicellular, diploid sporophyte. The sporophyte tissue undergoes meiosis to produce haploid spores. Haploid spores divide by mitosis to become multicellular, haploid gametophyte tissue. The gametophyte tissue undergoes mitosis to form haploid gametes. The complementary gametes fuse in fertilization.

o Vascular plants can either be **homosporous** or **heterosporous**. Homosporous plants produce one type of spore that can become either a male gametophyte (**antheridium**) or a female gametophyte (**archegonium**). Heterosporous plants produce two different types of spores. **Megaspores** always become archegonia. **Microspores** always become antheridia.

o **Seeds** are reproductive structures that greatly enhance reproduction in gymnosperms and angiosperms. Seeds have a tough seed coat that protects the embryo from dessication. The seed has food reserves called endosperm that nourishes the embryo while it remains in the seed. The seed allows for increased dispersal of embryo plants.

o Airborne pollen is a great advance in seed plants. The tough sporopollenin protects the sperm as they travel through air or by animal carriers to female gametophyte tissue. Each pollen grain contains two sperm cells.

o **Flowers** are an advanced reproductive structure in angiosperms. Flowers have separate male and female structures to enhance insect-mediated pollination. The **stamen** has a large bulbous end that is a mass of pollen called the **anther**. The **filament** anchors the anther to the flower. The **carpel** houses the **ovules** with eggs inside the ovary. The channel leading to the ovules is called the **style**. The sticky surface feature is the **stigma**. It is the site where pollen grains land. The brightly colored **petals** attract insects and animals to the flower. The **sepals** protect the floral parts as they develop.

o In angiosperms, the egg structure is unique. It divides to become an eight-nucleated cell. Then, that eight-nucleated entity is fertilized through a process called **double fertilization**. In this process, a pollen grain sticks to the stigma of the carpel. Enzymes in the stigma cause the pollen grain to form a pollen tube— a long extension for sperm travel. The two sperm travel down the pollen tube which extends through the style into an ovule at the base of the carpel. The two sperm cells from that single pollen grain go to the same egg in one of the ovules. One of the sperm cells fuses with one nucleus to become the embryo. The other sperm cell fuses with two polar nuclei to produce triploid endosperm (nutrient source inside the seed).

o Fruit forms around ovules with fertilized seeds. The fruit forms as the walls of the ovary swell. The fruit attracts animals who eat the fruit and disperse the seeds to new locations.

ANIMAL NUTRITION

• *Specialized structural compartments allow for digestion to take place.*

o **Intracellular digestion** occurs when food vacuoles fuse with lysosomes, mixing food with enzymes and allowing digestion to take place.

o **Extracellular digestion** involves the breakdown of food outside of cells, a much more complex form of digestion.

- *The structure of the teeth of mammals and other vertebrates greatly affects their diet.*
  - o **Carnivores** often have very sharp teeth used for ripping and shredding their prey.
  - o **Herbivores** have flat teeth more adapted for grinding the cellulose cell walls of the plant material thy have consumed.
  - o **Omnivores** have a combination of these types of teeth, as their diet consists of plant and animal material.

- *The length of the digestive system is directly proportion to the diet of vertebrates.*
  - o The length of the digestive system in omnivores, and especially herbivores, is generally longer than that of carnivores. Plants, having cells walls composed of cellulose, a relatively difficult to digest material, take much longer to break down. Therefore, a longer alimentary canal is necessary.

- *Diverse muscle structure throughout the esophagus allows for the transport of food from the oral cavity to the stomach.*
  - o Skeletal, or voluntary muscle, begins the process of swallowing, which is then followed by involuntary smooth muscle contraction.
  - o **Peristalsis**, or the alternation between a relaxed state and a contracted state propels food along the tract.

- *The epithelium of the stomach is covered with deep pits, increasing surface area and therefore the functional efficiency of the stomach.*
  - o By increasing the surface area of the stomach, more substances can be secreted for digestion by the gastric glands.
  - o These **gastric glands** are composed of three different types of cells: mucus cells, chief cell and parietal cells.
  - o **Mucus cells** secrete a lubricating mucus.
  - o **Chief cells** secrete an inactive enzyme called pepsinogen.
  - o **Parietal cells** secrete hydrochloric acid.

- *The structure of the small intestine provides for increased surface area, therefore increasing the efficiency of the absorption of nutrients.*
  - o **Villi**, tiny fingerlike projections, line the interior of the small intestine. These projections, along with **microvilli** allow nutrients to be absorbed at a much higher rate.

ANIMAL CIRCULATION (LAB #10) and ANIMAL RESPIRATION

- *Simple structural gastrovascular cavities are functionally sufficient for animals with a few cell layers.*
  - o As long as diffusion distances are short, gastrovascular cavities are adequate for internal transport. However, as the number of cell layers increases and diffusion distance increases, a circulatory system must be present to overcome this.
    - ▪ Circulatory systems have three main parts: fluid, vessels and a pump.
    - ▪ Circulatory systems may be **open** (interstitial fluid or hemolymph is continuous) or **closed** (there is a separation of blood from interstitial fluid).

- *The complexity of the structure of the organism greatly affects the efficiency of the metabolic rate of animals.*
  - o Animals with higher metabolic rates have more complex circulatory systems. As the number of circuits increases, blood can be delivered to the organs under higher pressure.
    - ▪ Two chambered heart: There is one atrium and one ventricle, with a single circuit of blood flow. Blood in the heart is always deoxygenated, and as blood leaves the ventricle it goes to the gills. Here, the blood exchanges oxygen for carbon dioxide and then travels to the systemic capillaries. Finally, the blood returns to the atrium of the heart.
    - ▪ Three chambered heart: There are two atriums and one ventricle, with two circuits of blood flow. As blood leaves the ventricle it goes through a dual artery that sends some blood to the lungs and skin and some blood to the body organs. Blood returns through veins to a respective atrium, where it enters the ventricle once again. In this system there is some mixing of oxygenated and deoxygenated blood.
    - ▪ Four chambered heart: There are two atriums, two ventricles and two circuits of blood flow. This system does not allow any mixing of deoxygenated and oxygenated blood. Blood leaves the left ventricle, travels to the body, returns to the right atrium, right ventricle, lungs, and left atrium, and left ventricle before being pumped to the body organs once again. This system allows for the greatest metabolic rate and is essential as animals became endothermic.

- *The functions of different types of blood vessels are directly related to their different structures.*
  - o The three main types of blood vessels are arteries, veins and capillaries.
  - o **Arteries** consist of three main layers of tissue, an outside layer of elastic, connective tissue, a middle layer of smooth muscle, and a smooth surface on the inside to minimize friction within the vessel. The outer and middle layers are very thick to accommodate the great pressure and speed with which blood travels through these vessels.

○ **Veins** consist of the same three layers, but have thinner middle and outer layers. This is due to the fact that blood travels through veins at a much reduced pressure and a much slower rate. However, in order to accommodate this, veins are surrounded by **skeletal muscle** and contain valves to prevent a backflow of blood.

○ **Capillaries** only consist of a very thin wall of endothelium, allowing for the efficient exchange of carbon dioxide and oxygen, as well as other substances.

• *Respiratory surfaces are usually thin and have large surface areas in order to increase the efficiency of gas exchange.*

○ The movement of oxygen and carbon dioxide across membranes is entirely passive, through the process of **diffusion**.

○ Endotherms will tend to have higher surface areas for gas exchange than an ectotherm of the same size.

○ Respiratory surfaces are moist to increase the respiratory rate.

○ In mammals, there are tiny air sacs called **alveoli**, where gas exchange occurs. This greatly increases surface area and the rate of diffusion.

• *The arrangement of capillaries within the respiratory beds of a fish gill increases the efficiency of gas exchange.*

○ Water and blood flow in opposite directions in this system, known as **countercurrent exchange**.

○ There is always a favorable gradient for the transfer of oxygen from the water into the blood along the entire length of the capillary.

ANIMAL REPRODUCTION

• *All animals have evolved advanced reproductive structures to increase overall reproductive fitness.*

○ Sexual reproduction occurs in all animals. Asexual reproductive mechanisms are possible in some animals.

○ **Budding** is a simple form of asexual reproduction where a new organism buds off of the side of an existing organism. The offspirng is a genetic clone of the parent offspring. It occurs in some cnidarians and poriferans.

○ **Parthenogenesis** is a form of asexual reproduction where eggs are made by meiosis. Then, the eggs begin cleavage and development without being fertilized. The offspring are haploid individuals. They are genetically unique due to crossing-over and independent assortment during meiosis.

○ If reproduction is sexual, the fertilization process can occur externally or internally. **External fertilization** happens in aquatic organisms that release sperm and eggs in the water where the complementary gametes fuse. This external fertilization occurs in mollusks, fish, amphibians, etc. **Internal fertilization** occurs when the male transfers sperm to the female's body and fertilization occurs within the female body. This process occurs in terrestrial organisms and many large aquatic organisms.

- *In mammals, fertilization is internal following sperm transfer to the female body through copulation. This process is due to advanced reproductive structures that are maintained hormonally.*

  o **Androgens** like testosterone maintain the male reproductive structures, stimulate secondary sex characteristics in males, and maintain male sex drive.
  o **Estrogen** maintains the reproductive structures of females and maintains secondary sex characteristics in females.
  o **Progesterone** maintains the lining of the female uterus so that if fertilization happens, a hospitable environment is prepared for the growing embryo.
  o The male reproductive system contains three sets of organs, the external organs, the duct system organs, and the accessory organs.
    ▪ One of the external organs is the **penis** which acts as the primary copulatory organ. It is designed for sperm transfer due to its rigidity that is reinforced by a bone in some mammals and by erectile tissue in other mammals. The other major external organ in males is the **scrotum** which holds the **testes** outside of the primary body cavity to maintain a steady temperature for sperm development.
    ▪ The duct system begins with the sperm producing structures: the testes. The testes have **seminiferous tubules** that undergo **spermatogenesis** to form haploid sperm cells. The sperm cells move to the epididymis where they mature. Then, during ejaculation they are forcefully expelled through the remainder of the male duct organs: the **vas deferens**, the **ejaculatory duct**, and the **urethra**.
    ▪ The male accessory organs include the **seminal vesicles**, the **prostate**, and the **bulbourethral gland**. The seminal vesicles and prostate gland add nourishing and lubricating materials to the sperm. These materials form the semen in which the sperm travels. The bulbourethral gland produces pre-ejaculatory fluid that neutralizes the acidity (due to being a common duct for urine and sperm in males) the urethra before the sperm travel through this structure.

- *The female reproductive system contains three sets of organs: the external organs, the duct system, and accessory glands.*
  o The external organs in females are housed in an area called the **vestibule**. It contains the vaginal opening which acts as the site of sperm deposition. Near that opening is erectile, stimulatory tissue called the **clitoris**. The **labia minora** are skin folds that cover and protect this area. Larger **labia majora** covers the entire vestibule area.

o The duct system begins with ovum producing tissue: the **ovaries**. In the ovaries, many **primary oocytes** are housed in undeveloped **follicles**. After puberty, one follicle a month develops. As the follicle develops, the primary oocyte continues meiosis to form a **secondary oocyte**. This secondary oocyte is expelled from the ovary during **ovulation**. A ciliated end of the oviducts draws the secondary oocyte into the **oviduct**. If fertilization happens, it occurs in the oviduct. The oviduct continues to the uterus. If fertilization occurred the secondary oocyte completes meiosis to become an egg and they begins development. The developing embryo implants into the uterine lining. If no fertilization occurs, the oocyte is shed form the unterus with its lining during menstruation through the vagina.

o The **Bartholin's gland** secretes vaginal lubricant to aid copulation.

o **Mammary glands** produce nourishment for the young that develops.

MUSCLE CONTRACTION

• *Due to the contractile nature of muscle cells, muscular structures are bundled fibers to increase strength.*

o Muscular tissue is the only type of tissue found in any type of organism that is capable of contraction. It is only present in animals.

o Muscle cells are packaged together in bundles to add strength to the muscular structures. These bundles of cells called fascicles are bundled together to form muscular tissue.

• *Each muscle cell is able to contract through a cellular cascade. This contraction process is described by the sliding filament theory.*

o In the **sliding filament theory**, muscle contraction begins with **acetylcholine binding** to ion-gated protein channels on the muscle cell surface, causing an **action potential**. This action potential stimulates the release of stored calcium ions from specialized regions of the endoplasmic reticulum called **terminal cisternae**.

o The **calcium ions** bind to a protein called **troponin**. Troponin is a small protein that is connected to a larger protein structure called **tropomyosin**. The troponin-tropomyosin complex covers the binding sites for myosin on the **actin** of the **thin filaments** in the muscle cell. When the calcium ions bind to the troponin, the binding causes a conformational change to the troponin and the tropomyosin which uncovers the myosin binding sites on the actin of the thin filaments.

o In the **thick filaments** of the muscle cell, there is a large protein structure of **myosin**. Each myosin strand has an enlarged end called the cross-bridge. When ATP binds to the myosin cross-bridge, it is activated. An activated **myosin cross-bridge** will bind to the exposed binding sites of the actin on the thin filament. Then, during the power stroke the energy from the ATP reactivates the myosin head.

o A new ATP molecule must bind to the cross-bridge for it to release the binding site and become reactivated.

o This binding and contraction continues until the original calcium channels close and calcium ions are returned to the terminal cisternae.

o Acetylcholinesterase stimulates this return of calcium ions by breaking and removing the aceytylcholine from the surface protein.

ANIMAL SKELETONS

• *Animal skeletons are support structures that provide body support, forms body shape, and serves as muscle attachment sites. These structures are integral for safety, support, and movement.*

  o There are three common skeleton types: hydrostatic skeleton, exoskeleton, and endoskeleton.

  o **Hydrostatic skeletons** are pressurized cavities filled with water. They are common in soft-bodied animals like earthworms.

  o **Exoskeletons** are enlarged external structures. They provide protection, but limit movement due to their weight. Exoskeletons can be made of calcium in the clams and chitons, and made of chitin in the insects and crustaceans.

  o **Endoskeletons** are lighter in weight and allow for muscles to attach on all sides and therefore provide for enhanced movement.

    ▪ Endoskeletons can be cartilaginous as in the sharks or calcified as the bony fish, birds, reptiles, and mammals.

    ▪ In calcified endoskeletons, specialized cells called **osteocytes** pull calcium from the bloodstream to ossify or harden the cartilage bone framework.

    ▪ In mammals, embryonic skeletons begin as cartilage, and are ossified during development.

MULTIPLE CHOICE QUESTIONS

1. Cells are recognized by other cells due to the presence of this molecule on the plasma membrane

    (A)   integral proteins
    (B)   peripheral proteins
    (C)   phosopholipids
    (D)   cholesterol
    (E)   glycoproteins

2. A cell that produces large amount of proteins for export would have a tremendous amount of the following organelle

    (A)   smooth endoplasmic reticulum
    (B)   mitochondria
    (C)   chloroplasts
    (D)   rough endoplasmic reticulum
    (E)   peroxisomes

For each of the following questions, use the following choices

    (A)   chloroplast
    (B)   mitochondria
    (C)   nucleus
    (D)   smooth endoplasmic reticulum
    (E)   ribosome

3. Site of cellular protein synthesis

4. Site for detoxification of drugs and lipid metabolism

5. Site of aerobic respiration

6. Contains DNA and thylakoids

7. **Not** a membrane bound organelle

8. A blood vessel designed structurally to overcome high pressure and velocity, while minimizing friction, is most likely a(n):

    (A)   artery
    (B)   vein
    (C)   capillary
    (D)   venule
    (E)   vena cava

9. The theory of countercurrent exchange ensures all of the following EXCEPT

    (A)   the opposite flow of blood in the gills maximizes oxygen uptake and carbon dioxide loss
    (B)   small arteries carrying cool blood inward are next to small veins carrying warm blood outward in endothermic fishes.
    (C)   the loop of Henle in the vertebrate kidney concentrates urine by reabsorbing solutes and water before excretion.
    (D)   the rate of heat exchange between an animal and its environment is increased.
    (E)   heat is concentrated within the bodies of birds due to the arrangement of blood vessels in their legs.

10. An organism who consumes a great deal of plant material would tend to have which of the following adaptations?

    (A)   Increased stomach volume for holding plant material
    (B)   Elongated cecum housing symbiotic bacteria
    (C)   Teeth modified for ripping and cutting
    (D)   Increased secretion of CCK to break down cellulose in the small intestine
    (E)   A shorter large intestine to allow for more efficient digestion.

11. Proteins are composed of a long chain of

    (A)   amino acids
    (B)   monosaccharides
    (C)   disaccharides
    (D)   nucleotides
    (E)   fatty acids

12. The two strands of DNA are held together through

    (A)   covalent bonding
    (B)   ionic bonding
    (C)   hydrogen bonding
    (D)   disulfide bridges
    (E)   Van der waal interactions

Use the following plant reproductive structures for answers to questions 13-16

    (A)   flower
    (B)   fruit
    (C)   antheridium
    (D)   archagonium
    (E)   pollen grain

13. Allows for increased seed dispersal by animals

14. Male gametophyte tissue

15. Produced from megaspores in heterosporous plants

16. Contains a generative cell

17. The hormone testosterone

    I.   maintains the male reproductive organs
    II.   is responsible for secondary sex characteristics in males
    III. regulates male sex drive

    (A)   I only
    (B)   II only
    (C)   I and II only
    (D)   I and III only
    (E)   I, II and III.

18. Genetically-unique, haploid eggs develop into haploid individuals in

    (A)   budding
    (B)   internal fertilization
    (C)   external fertilzation
    (D)   parthenogenesis
    (E)   segregation

19. In muscle cells, the release of calcium ions from sarcoplasmic reticulum causes

   (A)   acetycholine binding to surface proteins
   (B)   uncovering the myosin binding sites on the actin
   (C)   the influx of potassium
   (D)   the efflux of sodium ions
   (E)   activation of the myosin cross-bridges

20. The best example of a hydrostatic skeleton would be witnessed in a(n)

   (A)   gorilla
   (B)   frog
   (C)   clam
   (D)   snail
   (E)   earthworm

FREE RESPONSE QUESTIONS

1. Sex is an important process to maintain genetic variability within a population.

    a. **Describe** key structures that enhance sexual reproduction in
        i   flowering plants
        ii.   humans

    b. What primary differences occur in the sexual life cycles of these two distinct organisms?

2. The relationship of structure to function is one of the central themes to biology. For three of the following structures, briefly **describe** how the components of the structure relate to function.

    a. four chambered heart
    b. mammalian digestive tract
    c. mammalian nephron
    d. plant root systems
    e. cellular cytoskeleton

NO TESTING MATERIAL PRINTED ON THIS PAGE

GO ON TO THE NEXT PAGE

Theme Six

Regulation

ENZYMES

- *Enzymes regulate biological reactions by binding to one of the reactants in the reaction. That reactant is the substrate that binds to the enzyme at its active site.*
  - Biological reactions have high-energy requirements. The energy needed to begin a reaction is called the **activation energy**.
  - An enzyme helps to lower the activation energy so that biological reactions can proceed more quickly.
  - The activity of an enzyme is regulated by many factors: inhibitors, temperature, pH, etc.
  - **Competitive inhibitors** are molecules that compete with the substrate for the enzyme's active site. These molecules interfere with the substrate binding to the enzyme by physically blocking access to the active site. The presence of these inhibitors slows the enzyme-catalyzed reaction.
  - **Noncompetitive inhibitors** bind to the enzyme outside of the active site. This binding changes the shape of the enzyme and its active site so that it is difficult for the substrate to bind.
  - An increase in temperature causes the reaction rate to increase as the molecules move at an increased speed and there are more collisions between enzyme and substrate molecules. However, at some point, higher temperatures will cause the denaturation of the enzyme which will stop the reaction from occurring
  - Each enzyme has an optimum pH range where it is most efficient. Outside of that range, the enzyme becomes less efficient.
  - **Allosteric regulation** involves a molecule binding at a site outside of the active site to make the enzyme more receptive to substrate binding or to make the enzyme inactive by making the active site unavailable.

MEMBRANE TRANSPORT

- *The structure of the cell membrane results in selective permeability of the membrane.*
  - Many molecules can pass through the phospholipid bilayer of the cell due to their chemical properties and size.
  - Small **hydrophobic** molecules such as carbon dioxide and oxygen can pass easily through this membrane.
  - **Hydrophilic** molecules, although small, cannot pass easily through the membrane.
  - Hydrophilic, polar molecules pass through **transport proteins** along the surface of the cell membrane.

91

o Large molecules can enter or leave a cell through the processes of **endocytosis** or **exocytosis**, respectively.

  ▪ Specific types of endocytosis include **phagocytosis** (engulfing a solid particle), **pinocytosis** (engulfing liquids and dissolved particles) and **receptor mediated endocytosis** (receptors along the membrane regulate which molecules enter a cell.

o A different concentration of molecules inside the cell compared relative to the concentration of molecules outside the cell creates a **concentration gradient**.

  ▪ **Passive transport** occurs when molecules travel WITH the concentration gradient; that is, from an area of higher concentration to that of lower concentration.

  ▪ **Active transport** occurs when substances travel against the concentration gradient, this requires the expenditure of energy in the form of ATP.

o **Ion pumps** are often involved in the movement of ions across plasma membranes. These pumps tend to create voltage gradients, which allow the cell to store energy that can be used for other cellular processes.

CELL SIGNALING AND CELL CYCLE

• *Cellular processes are controlled by external factors binding to the cell's surface. This external binding triggers a cascade of chemical processes that lead to the desired cellular activity.*

  o **Signal transduction pathways** involve an intricate cascade of chemical responses that relay surface messages to the cell's interior to yield the appropriate response.

  o Each transduction pathway involves three stages: reception, transduction, and response.

  o In the reception stage, a chemical messenger, hormone, or external compound binds to a protein receptor on the cell's surface.

  o There are three common types of protein receptors: G-protein-linked receptors, tyrosine-kinase receptors, and ion channel receptors.

  o With a **G-protein-linked receptor**, there is a G protein that lies on the underside of large membrane-bound integral proteins. The integral protein crosses both phospholipid layers of the cell membrane. A chemical receptor or signal molecule binds to the receptor site of the protein. Then, the membrane-bound protein experiences a conformational change that activates the G-protein on its underside. The G-protein then begins a series of protein-activating events in the cell's interior.

- With a **tyrosine kinase receptor**, there are often multiple signal binding sites on the cell surface. There are often two large membrane proteins with each having its own binding site. When the signal binds, the two proteins come together through a conformational change and become a dimer. Then, the multiple tyrosine amino acids on each of the proteins become receptive to activation by ATP. Phosphates bind to the tyrosines to activate it. These phosphates are transferred to a relay protein to begin the cell's internal events.
- With an **ion channel receptor**, there is a large membrane-bound protein that can act as an ion channel. When a chemical signal binds to its surface, a conformational change allows the ion channel to open and many specific ions flow through the channel changing the concentration of that ion inside the cell. The electrical change inside the cell can trigger internal cellular activity.
- The **transduction** of the external signal is achieved mainly through a series of protein activations and inactivations.
- A **protein kinase** phosphorylates another protein to activate it. After the signal has been interpreted, an activated kinase begins a cascade of protein phophorylations. Each protein activates the next protein in the cascade.
- **Protein phosphatases** remove the phosphates from proteins and deactivate proteins.
- In addition to proteins, there are non-protein members of the internal cascade called **second messengers**. Cyclic AMP, calcium ions, inositol triphosphate and diacylglycerol are common second messengers.
- After the cascade of protein activations and second messenger activity, the actual cellular response is achieved.
- The cascade allows for regulation and amplification of the original signal. One signal type can yield different responses in different cell types. One signal molecule can yield a large-scale response due to amplification as the message passes through the cascade.

- *The cell cycle is controlled by molecules that affect growth, development and maintenance by regulating the frequency of cell division.*
    - There is a critical control point in the cell cycle called a checkpoint. If certain cellular processes have been completed, the cell cycle will continue. There are also outside signals to determine whether or not the process will proceed.
        - The G_1 checkpoint is critical in the cell cycle. If a cell does not pass the G_1 checkpoint, it will proceed into the G_0 state and become a nondividing cell.
        - The main molecules that fluctuate in a cell to regulate the activity of the cell cycle are **cyclins** and **kinases**.
        - **Cyclins** accumulate in the cell and combine with **cyclin dependent kinases**. This combination results in molecules of **MPF**, which allow the cell to pass the G_2 checkpoint.

■ **MPF** levels peak during metaphase; during anaphase, the cyclin within the MPF is broken down and the cell enters the G1 phase.

○ **Growth factors** are also released and stimulate other cells to divide.
○ **PDGF** is a growth factor released by the platelets in which stimulates fibroblast division. PDGF binds to the receptors on the fibroblast and allows the cell to pass the G1 checkpoint

- *Physical factors have an effect on cell division. Cancer cells often lack one or both of these control factors.*
  ○ **Density dependent inhibition** describes how cells will stop dividing as a cell population reaches a specific density.
  ○ **Anchorage dependence** describes how cells must be attached to a particular substrate in order to divide.

GENOME REGULATION

- *Bacterial genes are regulated through one primary process: an operon. An operon consists of a DNA sequence close to the promoter acting as an operator. Binding of repressors at this site help regulate binding of RNA polymerase to the promoter.*
  ○ Repressor proteins bind to the operator sequence on the genome. If a repressor is present, the RNA polymerase cannot bind to the promoter sequence and begin transcription. If a repressor is absent, the RNA polymerase will bind to the promoter to begin transcription.
  ○ Genes with repressors bound to the operator site are inactive. Genes lacking repressors bound to the operator site are active.
  ○ **Repressible operons** generally lack repressor proteins and precede gene systems that are normally active. The tryptophan operon is an example. When excess tryptophan is present it acts as a corepressor. The tryptophan binds to the repressor causing a conformational change that makes the repressor bind to the operator inactivating the gene system. When the corepressor is not present in excess, the repressor does not bind to the operator.
  ○ The lactose operon is an example of an **inducible operon**. A gene regulated by an inducible operon is generally inactive due to a repressor being bound to the operator. These gene systems only become active when an inducer (allolactose in this case) binds to the repressor and prevents it from binding with the operator. When the inducer is absent, the repressor returns to the operator to inactivate the gene system.

- *Many pre- and post-transcriptional processes regulate eukaryotic genes. These processes can prevent transcription, increase the rate of transcription, prevent translation, or interfere with final protein processing.*
  ○ Since all cells in a eukaryotic organism have the same genetic information, the level of gene activation determines the specific cell type and allows for differing cell types.

Regulatory Process	Effect on Protein Synthesis	Description of Process
Chromosome compaction	Prevent transcription	Pieces of chromosome remain as heterochromatin (compact) that makes the promoter sites inaccessible to the RNA polymerase.
DNA methylation	Prevent transcription	Methyl groups attach to the DNA nucleotides blocking the binding of the RNA polymerase to the promoter.
Histone acetylation	Increase rate of transcription	Acetyl groups are bound to the histone proteins in the chromosomes. This binding causes the histones to loosen their grip on the DNA. The DNA coils less tightly making it easier for the RNA polymerase to bind to the promoter.
Control Elements	Alter transcription rate (increase or decrease)	Genome sequences serve as protein binding sites. Some proteins act as activators and encourage RNA polymerase binding. Other proteins act as silencers and interfere with the binding of the RNA polymerase.
mRNA end cap removal	Prevent translation	The poly-A tail or modified guanine cap on the mRNA can be removed. Then, the mRNA is degraded before it is translated.
Alternative splicing	Alter protein product	An RNA and protein containing spliceosome complex recognizes different sequences as introns in the mRNA. Then, the mRNA sequence can vary and a different protein can be produced from the same original gene sequence.
Interference with protein-folding	Prevent functional protein product	Chaperone proteins are obstructed to interfere with the folding process. Unfolded proteins are non-functional.
Protein degradation	Remove protein excesses	Excess proteins are tagged with ubiquitin markers. Then, a proteosome complex recognizes the tags and destroys the excess proteins.

PLANT HORMONES

Plant Hormone	Primary Functions
Auxin	Stimulates cell elongation, concentrated in growth regions of the plant (stem and root tips), responsible for phototropic and gravitropic growth responses, causes apical dominance
Cytokinin	Stimulates cell division, concentrated in growth areas of the plant
Gibberellin	Breaks seed dormancy, responsible for rapid bursts of growth, stimulates division and cell growth
Abscisic acid	Maintains seed dormancy, inhibits growth
Ethylene	Gaseous hormone that inhibits growth, stimulates fruit ripening
Brassinosteroid	Inhibits growth, slows leaf abscission

• *Chemical messengers often regulate the physiological processes in plants. These hormone molecules regulate major growth activities in the plant as they move from cell to cell through the plant tissue.*

CHEMICAL MESSENGERS IN ANIMALS

• *Chemical messengers regulate many long-term body processes in animals. Chemical messengers bind to surface proteins on a target cell and invoke a transduction pathway that leads to the desired cellular response.*
  o In animals, there are two distinct classes of chemical messengers: local regulators and long-distance chemical messengers.
  o The two types of local regulators are paracrine signals and synaptic signals.
    ▪ **Paracrine signals** are chemical messengers released from cells that are neighbors to the target cell. After being released into the environment by exocytosis, the paracrine signal binds to receptors on adjacent cells to initiate a response.

    ▪ **Synaptic signals** involve the release of neurotransmitters through exocytosis from the axon terminals of a neuron. The neurotransmitter crosses a gap called a synapse to receptor sites on proteins on the target cell surface. This surface binding initiates a cellular cascade that will yield the desired response.

    ▪ **Long-distance signaling** in animals is through the production of hormones. Hormones are chemical messengers that travel through the bloodstream to reach target cells that can be very far away from the gland of release.

  o Hormones are secreted by endocrine organs. This system of organs includes the hypothalamus, pituitary gland, pineal gland, thyroid gland, thymus gland, parathyroid glands, pancreas, adrenal glands, and gonads.

Hormone	Gland of Release	Actions of Hormone
Inhibiting hormone	Hypothalamus	Inhibits the release of hormones from the anterior pituitary gland
Releasing Hormone	Hypothalamus	Stimulates the release of hormones from the anterior pituitary gland
Antidiuretic hormone	Posterior Pituitary	Stimulates the nephrons of the kidneys to reabsorb more water
Oxytocin	Posterior Pituitary	Stimulates uterine contractions; promotes milk ejection from mammary glands
Prolactin	Anterior Pituitary	Stimulates milk production in mammary glands
Follicle-stimulating hormone	Anterior Pituitary	Stimulates sperm and egg production
Luteinizing hormone	Anterior Pituitary	Stimulates maturation and development of testes and ovaries
Thyroid-stimulating hormone	Anterior Pituitary	Stimulates release of hormones from the thyroid gland
Adrenocorticotropic hormone	Anterior Pituitary	Stimulates the release of glucocorticoids from the adrenal cortex
Growth hormone	Anterior Pituitary	Stimulates bone growth and cellular growth
Melatonin	Pineal gland	Related to mood due to seasons and maintains biological rhythms
Calcitonin	Thyroid gland	Reduces blood calcium levels
Parathyroid hormone	Parathyroid gland	Increases blood calcium levels
Insulin	Pancreas	Reduces blood glucose levels
Glucagon	Pancreas	Increases blood glucose levels
Epinephrine and norepinephrine	Adrenal medulla	Regulate nervous functions, involved in maintaining metabolism and constricting blood vessels
Glucocorticoids	Adrenal cortex	Increases blood glucose levels
Mineralocorticoids	Adrenal cortex	Increases absorption of Na+ in the nephrons of the kidneys
Androgen (testosterone)	Testes	Maintains male reproductive organs, maintains secondary sex characteristics in males, sex drive
Estrogen	Ovaries	Development of female reproductive organs and secondary sex characteristics in females
Progesterone	Ovaries	Promotes growth of the uterine lining

HOMEOSTASIS

- *Maintaining temperature is very important in order to regulate biochemical processes.*
  - **Conduction, convection, radiation,** and **evaporation** are the main ways an organism gains and loses heat to/from the environment.
  - **Endotherms** and **ectotherms** manage their heating expenses differently. Endotherms need to work to generate heat to keep it at a specific temperature; ectotherms maintain body temperatures close to their surrounding environment.
  - **Endotherms** have many advantages. Due to the constant body temperature, specific metabolic processes are continuous, extreme exercise is possible and animals can live in very low temperature habitats. However, these advantages require great energy expenditure.

- *Endotherms and ectotherms manage heating expenses differently and use different means to thermoregulate.*
  - Many endotherms use adjustments in blood flow to alter temperature.
    - **Vasodilation** (increase of the diameter of blood vessels) and **vasoconstriction** (decrease of the diameter of blood vessels) are used to warm or cool the tissues, respectively.

    - **Countercurrent exchange mechanisms** are also used to regulate temperature. The arrangement between arteries and veins, both carrying blood of differing temperatures, result in an efficient heat exchange. Some ectotherms such as large fish (tuna) also utilize this to retain heat in the central core of their body.

    - Organisms may also use evaporative cooling to lower their body temperature.

    - Many birds and mammals also use hair or feathers to insulate.

    - **Torpor** can also be used to manage temperature needs.
  - **Hibernation,** or long-term torpor, causes body temperature and metabolic rate to decrease dramatically.
  - **Estivation** or summer torpor is like hibernation in the sense that there is a decrease in body temperature and metabolic rate, but in this instance, it is in response to high temperatures and lack of water.
  - Ectotherms tend to use behavioral means to adjust their temperatures.
    - Many animals seek out or avoid warm areas, depending on their needs.

EXCRETION

- *Osmoregulation is necessary to balance the control of solutes between interstitial fluid and the external environment. Transport epithelia are essential in regulating water balance.*

- *Osmoregulators also produce different types of nitrogenous wastes in order to maximize the efficiency of their excretory system.*
    - o **Ammonia** is relatively toxic to terrestrial animals and requires large amounts of water to dilute it, and is therefore the main nitrogenous waste for aquatic animals. It does not require a lot of energy to produce and is often secreted across the gills.
    - o **Urea** takes more energy to produce but it is more tolerable by terrestrial animals. The liver makes the urea, and because of its relatively low toxicity, allows the animal to conserve more water than with the excretion of ammonia.
    - o Terrestrial animals and birds have water conservation concerns and therefore some secrete a dry nitrogenous waste called **uric acid**. While it is not soluble in water, and thus a great solution for animals that have little access to water, it requires the most energy to produce.

- *Mammalian kidneys have structural adaptations to allow for the regulation of water balance.*
    - o The kidney consists of about a million nephrons that **filter, reabsorb, secrete, and excrete** substances from the body.
    - o **Filtration** occurs as blood is forced from the glomerulus to the Bowman's capsule. Small molecules can pass through this filter, but it is very nonselective and many molecules that the body still wants to maintain are forced across the membrane in this process.
    - o **Reabsorption** occurs throughout the proximal tubule, loop of Henle and distal convoluted tubule in the nephron, reclaiming NaCl, nutrients, water, potassium ions, and bicarbonate, which were originally lost from the blood stream at the Bowman's capsule.
    - o **Secretion** occurs throughout the nephron, letting molecules that may not be in the filtrate at this time, be excreted. These are often large molecules or waste products, from the liver that could not pass across the membrane in the Bowman's capsule.
    - o **Excretion** occurs at the collecting duct, producing relatively concentrated urine, conserving as much water as possible, reclaiming nutrients and specific ions, and ridding the body of nitrogenous waste.

- *The regulation of water balance is addressed by adaptations within the excretory system of vertebrates.*
  - The main site for water reabsorption within the vertebrate kidney is in the **loop of Henle**.
  - The longer the loop of Henle is, the greater the conservation of water. This conservation will maintain a steep osmotic gradient and produces extremely concentrated urine which is a useful adaptation for organisms in dry, arid environment.

- *Many hormones regulate water conservation and kidney function.*
  - **Antidiuretic hormone (ADH)** is produced in the hypothalamus and is released when the osmolarity of the blood increases beyond a specific point. At this time, ADH targets the distal tubules and collecting ducts of the nephron, increasing their permeability to water, increasing water conservation.
    - Alcohol and caffeine can interfere with this hormone by inhibiting its release.
  - The RAAS (renin-angiotensin-aldosterone system) also regulates kidney function. As blood pressure and volume drop, renin is released, converting angiotensinogen (an inactive precursor) to angiotensin. This causes an increase in blood pressure and in blood volume by decreasing blood flow to capillaries, constricting arterioles and stimulating the reabsorption of NaCl and water. It also stimulates the release of aldosterone causing the nephron to reabsorb more sodium and water.
  - **Atrial natriuretic factor** (ANF) works in opposition to the RAAS. Instead of responding to a drop in blood pressure, ANF causes an increase of blood pressure. ANF also inhibits NaCl reabsorption and the release of aldosterone.

NERVOUS SYSTEM

- *The nervous system integrates a great amount of information and regulates responses to this information.*
  - Input into the nervous system occurs at a **sensory receptor** and is conveyed to the central nervous system (CNS). The CNS consists of the brain and spinal cord and integrates the information, forming a response. This response is transmitted to **effector** cells (muscles or glands).

- *Nerve signals are regulated by the maintenance of membrane potentials across a plasma membrane.*
  - A concentration gradient exists between the inside and outside of a nerve cell. Specific solutes (anions and cations) cause the environment to be electrically charged.
    - The inside of the neuron is highly concentrated with anions.

■ The outside of the neuron is highly concentrated with cations.

■ This distribution creates a **resting membrane potential** in the neuron of about –70mV. This potential is maintained through the use of the sodium-potassium pump that uses ATP to actively pump Na+ out of the cell, and K+ into the cell.

• *If the membrane potential changes, an electrical nerve impulse will be generated.*
   o Excitable cells can cause changes in the membrane potential of a neuron, as they contain **ion channels** that allow for the movement of cations and anions within the neuron.
   o As an influx of Na+ occurs in an excitable cell, the interior of the neuron becomes more positive. If the cell becomes positive enough (-50 mV), it reaches **threshold**, additional Na+ gates open and the neuron will fire. This is called **depolarization**, or a reduction in the voltage across a membrane.
   o As the voltage reaches about +35 mV, there is a closing of Na+ gates and K+ gates open. Potassium ions rush out of the cell, making the neuron more negative, and returning the neuron to its initial state.
   o There is an **undershoot** (-75mV) where K+ gates remain open too long. This prevents the signal from reversing back along the neuron.
   o The sodium potassium pump returns the membrane to its original resting potential. It takes only a millisecond to return the neuron to resting state, and the neuron may fire again.

• *Chemical transmission regulates communication between neurons.*
   o **Synapses** are spaces between neurons. As the electrical impulse reaches the synaptic terminal, a chemical known as a **neurotransmitter** is necessary in order to stimulate an adjacent neuron.
   o As a synaptic terminal becomes depolarized, it triggers a release of Ca+ ions within the membrane. This causes synaptic vesicles, containing neurotransmitters, to bond with the **presynaptic membrane** and diffuse across the space to the **postsynaptic membrane**. The neurotransmitter molecules are taken up by the adjacent neuron causing an action potential to occur in the next neuron when the threshold is reached.

• *Regulation in the nervous system is due to the release of different neurotransmitters on different types of cells.*
   o Different receptors on postsynaptic cells affect the action of the neurotransmitter.
   o Some neurotransmitters are excitatory, some are inhibitory and some can be both.
   o Some neurotransmitters are long acting, some are quick acting.
   o Secretion sites (CNS or PNS) can vary for neurotransmitters.
   o Some of the main neurotransmitters are: acetylcholine, norepinephrine, dopamine, serotonin, some amino acids, and neuropeptides.

- *The division of the vertebrate nervous system into a central and peripheral branch, which is further divided, allows for the maintenance of homeostasis.*
  - o The peripheral nervous system (PNS) is divided into a sensory and motor branch. The motor division is further divided into an autonomic (automatic) branch and a somatic (voluntary) branch. The autonomic branch is finally divided into parasympathetic and sympathetic divisions.
- *The sympathetic division is activated during times of stress.*
  - o Heart rate and gas exchange increase.
  - o Digestion is slowed.
  - o Adrenaline is secreted.
- *The parasympathetic division is activated during times of relaxation.*
  - o Heart rate decreases.
  - o Digestion is enhanced.
- *Circadian rhythms regulate many complex processes.*
  - o These processes include sleep/wake cycles, hormone release and hunger.
  - o Structurally, this control is usually present in the hypothalamus of the brain of complex organisms. It may also be present in other parts of the boy, such as the wings of fruit flies.
  - o Biological clocks require external cues in order to function correctly.

IMMUNITY

- *The immune system in animals strives to free the body of harmful pathogens that try to enter through body openings. This system regulates blood cell action and body responses to these foreign pathogens.*
  - o There are nonspecific and specific immune responses. **Nonspecific immune responses** combat any foreign pathogen and **specific immune responses** only react to a specified antigen.
  - o There are two lines of nonspecific immunity: barrier mechanisms and phagocytic cells.
    - ▪ **Barrier mechanisms** block or filter any foreign antigen from infecting the body. Barrier mechanisms include the skin, mucous, hairs, and antimicrobial proteins. The skin is composed of tightly-packed epithelial cells that block entrance of microorganisms. The skin has an acidic pH that is not desirable to microorganisms. The mucous and hairs trap dirt and microbes. Antimicrobial proteins are found in tears, saliva, and lining body cavities. These proteins cause the lysis of bacterial cells by cell-membrane rupture.

    - ▪ **Phagocytic cells** are also nonspecific. These cells use phagocytosis to ingest foreign cells. Many white blood cells like the monocytes, basophils, and eosinophils are phagocytic defense mechanisms.

o In specific immunity, there are two common processes: **cell-mediated** or **humoral immunity**.

o Both specific responses begin with antigen-presenting cells and helper T cells. A phagocytic cell engulfs the foreign body and becomes an antigen-presenting cell. The antigen-presenting cell causes T cell proliferation to form activated helper T cells and memory helper T cells. The activated helper T cells have a surface protein complex called MHC I with a specific protein called CD4 that serves as the binding site for the antigen-presenting cell. The activated helper T cells secrete cytokines that serve as chemical messages that promote cell-mediated and humoral responses.

▪ In the **cell-mediated response**, the cytokines from the helper T cell signal the cytotoxic T cell to bind to the target cell at its MHC-I protein complex through the CD8 protein. Upon binding, the cytotoxic T cell secretes perforin and other enzymes that rupture and destroy the target cell.

▪ In the **humoral response**, the cytokines from the helper T trigger the B cell to bind to the foreign antigen. The B cells proliferate and form plasma cells and memory B cells. Plasma cells produce and secrete antibodies that bind to the foreign antigen and block its activity. The memory B cells remain in the bloodstream to greatly increase the rate of attack against the same pathogen.

o Second exposure to the same antigen results in a faster response rate due to the memory helper T cells and memory B cells in the bloodstream that allow for more efficient activation of the immune response.

ECOLOGY

• *Population growth is regulated by interactions and the transfer of energy within ecosystems.*

o **Density dependent factors**, including a limited amount of food or space, increased predation, disease and stress regulate population size.

o **Density independent factors**, including weather related events, may regulate population size.

o Many populations have regular predictable **boom and bust cycles** in response to seasonal food shortages, predator prey interactions, or a combination of the two.

o The transfer of energy provides for a stable ecosystem.

▪ The loss of energy at each trophic level is between 80% and 95%, depending on the type of ecosystem.

▪ This transfer of energy only provides for a specific number of individual organisms (and biomass) in each trophic level to be sustained. This regulates the population size and determines the **carrying capacity** (the maximum population size allowed by resources) of an ecosystem.

PLANTS

- ***The opening and closing of stomata regulate water loss within plants.***
  - The surface of a leaf is covered with pores called **stomata**. These pores are surrounded by cells called **guard cells** that control the diameter of the stomata, regulating water loss.
  - **Guard cells** push outward when they are turgid. This turgidity is caused by an influx of K+ into the guard cell creating a hypertonic environment. Water from the surrounding area enters through osmosis, opening the stomata. As the K+ leaves the guard cells, the stomata close and prevent water loss.
  - **Aquaporins** may be involved in this process by affecting the permeability of the membrane to water.
  - Light stimulates the guard cells to accumulate K+ due to the blue light receptors in the membranes of the guard cells. This activates the ATP proton pumps in the membrane and encourages the influx of K+.
  - Internal clocks also affect stomatal openings. Plants left in the dark often continue to open and close their stomata on a 24 hour cycle.

- ***Adaptations for photosynthesis regulate the production of sugar and maximize photosynthetic efficiency in plants.***
  - C_4 plants increase photosynthetic efficiency by altering their mode of carbon fixation. They incorporate the carbon dioxide into a four carbon sugar in the mesophyll cells of a leaf and then export those four carbon sugars to the bundle sheath cells. This minimizes **photorespiration** (oxygen is fixed and no ATP is produced).
  - **CAM** plants open their stomata at night and store carbon dioxide in organic acids until the day. Then, these organic acids are converted back to carbon dioxide for the Calvin cycle while the stomata are closed. This minimizes water loss and maximizes photosynthesis.

MULTIPLE CHOICE QUESTIONS

1. Which statement is true as to how enzymes affect the activation energy of a chemical reaction?

   (A)   Enzymes provide the activation energy for the reaction.
   (B)   Enzyme binding replaces the energy demands of a reaction.
   (C)   An enzyme reduces the activation energy of the reaction.
   (D)   Enzymes increase the rate of chemical reactions without affecting the activation energy of the reaction.
   (E)   Enzymes inhibit the ability of a reaction to reach the energy barrier.

2. Which type of cell mechanism utilizes a protein that lies on the inside of the cell membrane that can become activated and deactivated to begin a signal cascade?

   (A)   Sodium-potassium pump
   (B)   Proton pump
   (C)   Ion channel receptor
   (D)   Tyrosine-kinase receptor
   (E)   G-protein-linked receptor

3. A signal transduction pathway helps to

   (A)   promote binding between chemical signals and receptor proteins
   (B)   amplify the cellular response from a small amount of signal
   (C)   signal transport of other materials across the cell membrane
   (D)   interfere with inappropriate hormonal signals
   (E)   increase a cell's receptiveness to new chemical signals

4. All of the following reduce the rate of gene activity in a eukaryote EXCEPT for

   (A)   histone acetylation
   (B)   DNA methylation
   (C)   heterochromatin areas of DNA
   (D)   silencers binding to control elements
   (E)   mRNA end cap removal

5. Alternative splicing describes a process where

   (A)   mRNA end caps are removed
   (B)   different DNA fragments are fused together by DNA ligase
   (C)   different regions of the RNA transcript are removed as introns yielding different mRNA strands
   (D)   amino acids are removed from growing polypeptides
   (E)   small amino acid fragments are fused together to yield quaternary protein structure

6. The plant hormone that is most responsible for the phototropic and gravitropic growth response causes cell elongation in growth regions of the plant. This hormone is

   (A)   gibberellin
   (B)   abscisic acid
   (C)   cytokinin
   (D)   auxin
   (E)   ethylene

7. A gaseous plant hormone that causes fruit ripening is

   (A)   Gibberellin
   (B)   Abscisic acid
   (C)   Cytokinin
   (D)   Auxin
   (E)   Ethylene

8. A hormone is a(n)

   (A)   long-distance regulator that travels through the blood stream
   (B)   synaptic signal
   (C)   paracrine signal
   (D)   local regulator that targets organs adjacent to the release site
   (E)   animal specific molecule

9. All of the following are examples of nonspecific immune defenses EXCEPT

   (A)   skin and barrier membranes
   (B)   lysozymes and antimicrobial proteins
   (C)   phagocytic cells
   (D)   antibody binding
   (E)   mucous and hair traps and filters

10. Helper T cells are involved with the

      I. Cell-mediated response
      II. Humoral response
      III. Nonspecific defenses

  (A) I only
  (B) II only
  (C) I and II only
  (D) II and III only
  (E) I, II, and III

Questions 11-14 refer to the following structures of the mammalian kidney.

  (A) glomerulus
  (B) distil tubule
  (C) loop of Henle
  (D) proximal tubule

11. First site of reabsorption of NaCl and water.

12. Aldosterone stimulates these cells to actively absorb sodium.

13. The length of this section varies with the water regulation needs of the organism.

14. Most reabsorption of water occurs here.

15. Which of the following choices maintain the voltage difference across the plasma membrane of the neuron?

      I. The nature of the lipid bilayer
      II. Transport proteins
      III. Concentration gradients

  (A) I only
  (B) II only
  (C) II and III only
  (D) I and III only
  (E) I, II and III

16. In a negative feedback system,

   (A)   a stimulus maximizes the body's response beyond a set point
   (B)   the effector leads the stimulus to change
   (C)   a stimulus causes the body's response to fall below a set point
   (D)   the receptor is often a muscle or a gland
   (E)   a stimulus causes a response that returns body levels to a normal level

17. All of the following processes are examples of mechanisms that do not require energy expenditure in cells EXCEPT

   (A)   osmosis
   (B)   diffusion
   (C)   endocytosis
   (D)   facilitated diffusion
   (E)   passive transport

Questions 18-20 refer to the following relative concentrations of solutes in a solution.

   (A)   isotonic
   (B)   hypertonic
   (C)   hypotonic

18. Placing a young plant cell in this environment could cause plasmolysis.

19. Placing red blood cells in this environment will cause them to burst.

20. A cell placed in this environment will cause a net influx of water into the cell.

FREE RESPONSE QUESTIONS

1. In humans, many body processes are regulated through the actions of hormones. In many cases, there are a pair of hormones that act antagonistically to regulate a given body process. Select **two** of the following four pairs of antagonistic hormones.

   a. Insulin and Glucagon
   b. Calcitonin and Parathyroid hormone
   c. Stimulating hormones and Inhibiting hormones from the hypothalamus
   d. Antidiuretic hormone and Mineralocorticoids

For each pair:

   a. **Describe** the action of each hormone
   b. **Describe** the gland of release of each hormone
   c. **Explain** how the hormonal regulation of that process is an example of positive or negative feedback.

2. The particular type of nitrogenous waste an animal excretes depends on evolutionary history and habitat, however the byproducts of metabolism are naturally very toxic, nitrogen containing wastes.

   a. **Identify** the three main types of nitrogenous waste and discuss their biochemical properties.
   b. **Explain** how habitat drives the evolutionary effectiveness of each type of nitrogenous waste.

NO TESTING MATERIAL PRINTED ON THIS PAGE

GO ON TO THE NEXT PAGE

Theme Seven

Interdependence in Nature

PHOTOSYNTHESIS AND CELLULAR RESPIRATION

- *Photosynthesis and cellular respiration are interdependent processes. The oxygen gas and glucose that are produced in photosynthesis are the reactants of cellular respiration and the carbon dioxide produced in cellular respiration is one of the reactants of photosynthesis.*

  o In the process of photosynthesis, water and carbon dioxide are used to form glucose and oxygen gas.
  o Oxygen gas is formed as water molecules are split to replace electrons in noncyclic photophosphorylation.
  o Glucose is formed during the sugar production stage of the Calvin-Benson cycle as carbon dioxides are fixed to existing carbon molecules by the enzyme rubisco.
  o The oxygen gas produced from photosynthesis is important for all organisms because it acts as the final electron acceptor in the electron transport chain of cellular respiration.
  o Glucose is the primary starting material for glycolysis.
  o During the pyruvate oxidation stage and the Krebs cycle of cellular respiration, carbon dioxide is produced. This carbon dioxide is one of the reactants that plants and other autotrophs need for photosynthesis.

EVOLUTION OF ORGANISMS

- *There is a diverse array of modern organisms. All of these organisms evolved from a common ancestor through similar evolutionary processes.*

  o Primary modes of evolution include: natural selection, non-random mating, gene flow, genetic drift, and mutation.
  o **Natural selection** describes a process where the organisms with the most advantageous environments in a given area will be able to have the greatest reproductive success.
  o **Directional selection** demonstrates a situation where one extreme phenotype is favored in the environment and one extreme phenotype is selected against.
  o In **stabilizing selection**, the intermediate phenotypes are favored while both extreme phenotypes are selected against.
  o In **diversifying selection**, the intermediate phenotypes are selected against and both extreme phenotypes are favored.

111

○ **Sexual selection** occurs when mating is not completely random. Some trait or behavior makes some members of a population more favorable mates for the opposite gender.

○ **Genetic drift** occurs when chance alone affects the frequency of alleles in a population. It can be seen in the founder effect or when the population size falls due to the bottleneck effect.

○ **Gene flow** occurs as new alleles are continually added and old alleles are removed from a population.

○ **Mutation** allows for the emergence of new alleles. It is a dramatic force in the evolution of new and novel traits.

○ **Convergent evolution** occurs when two populations or species of dissimilar organisms evolve similar morphological traits due to common selection pressures. These traits are analogous.

○ **Divergent evolution** occurs when evolutionarily-similar organisms evolve dissimilar traits due to different selection pressures. These traits are homologous.

POPULATION ECOLOGY

• *All members of a population are affected by common biological and geological factors.*

○ A **population** is a group of organisms of the same species living in the same area at the same time.

○ Populations are affected by **biotic** and **abiotic factors**. Biotic factors include interactions with all other types of living organisms whereas abiotic factors include non-living factors like climate, temperature, water availability, soil conditions, etc.

○ If the resources for a population are unlimited, the population will grow **exponentially**.

○ If the population size gets large and the population persists in a given environment for a long time, the resources will eventually become limited. These populations experience **logistic growth**.

○ The **carrying capacity** is the maximum population size that can be sustained in a given environment.

○ Opportunistic species called "r species" often exhibit exponential growth. Common traits of these species include: little or no parental care, semelparous, high fecundity, poor competitors, short life span, and short generation time. They are organisms that quickly move into open environments and flourish. However, more stable competitors eventually move in and displace these opportunistic species.

○ Stable species called "K species" often exhibit logistic growth. Common traits of these organisms are long life span, iteroparous, low fecundity, long generation time, good competitors, and parental care.

COMMUNITY ECOLOGY

- *Competition will occur between organisms that have similar lifestyles, requirements and are in the same environment.*

  o Competition is the interaction between two organisms for food, water, light, habitat, or mates. This interaction may be between organisms of the same species (**intraspecific competition**) or different species (**interspecific competition**).

    ▪ **Predator:** An organism that eats other organisms.
    ▪ **Prey:** An organism that is eaten by other organisms.

  o If the competition between two species continues, one species will likely become extremely efficient at obtaining food, water, light, habitat or a mate, and will eventually cause the extinction of the other species, a concept called **competitive exclusion**.
  o However, often times when similar species exist in the same environment, competitive exclusion does not occur. This is due to the fact that while they may live in the same environment, their **niche** may be different.
  o This niche is the role the species plays in the environment, and while some of the niche overlaps with another species, there are also differences in each **niche**.
  o Those organisms that have a great deal of niche overlap may actually be selected against, causing the species to diverge, and therefore minimize the overlap of the niche, promoting the success of each species.

- *The symbiotic relationship between two closely related organisms is a complex interaction that can result in harm or benefit to each respective organism.*

  o **Parasitism** occurs when one species benefits in the relationship and the other is harmed.
  o **Mutualism** occurs when both species receive benefits from the relationship.
  o **Commensalism** occurs when one species benefits in the relationship and the other is neither harmed nor helped.

ANIMAL BEHAVIOR

- *The environment and the genetic makeup of an organism play a large role in specific behaviors.*

  o Communication signals are important in the relationships between organisms.

    ▪ **Pheromones** are chemical substances that are released by organisms that affect the behavior of another organism.

- Organisms must be able to synthesize many types of pheromones for different messages.

  - **Courtship displays** are important in establishing a highly successful species. However, they are expensive to produce and hard to maintain. While they have survival costs they also have reproductive benefits and occur more often in males than in females.

- *Nature (innate behaviors) and nurture (learned behaviors) are both influential to the display of complex behaviors.*

  o Instinctive behaviors
  - A fixed action pattern is a behavior that is triggered by a specific stimuli and runs to completion regardless of environmental changes.

    - Once an organism encounters a specific **sign stimulus**, it will stimulate a **releasing mechanism** within the brain to stimulate a **releasing mechanism**.
    - Example: A goose retrieves an egg that has rolled from the nest. In this example, the fixed action pattern is the rolling of the egg back to the nest; the sign stimulus is the appearance of an object near the nest. If the object is not an egg, most likely the goose will keep rolling it, but not keep it. If the goose loses the egg during the process, it continues the motion of retrieval and needs to sit before it notices another egg and begins another fixed action pattern.

  - Kineses and taxes are changes in movements due to a particular stimulus.

    - **Kinesis** involves a random movement in response to a stimulus where a favorable environment is often encountered, but only by chance.
    - **Taxis** involves a directed movement in response to a stimulus to encounter a favorable environment.

  - Reflexes are instinctive movements in organisms.

    - **Reflexes** are very fast responses to a stimulus that often prevent unfortunate events. A motor neuron in the spinal cord directs the response, which does not require a motor response from the brain.

  o Learned behaviors
  - There is a critical period during development that can impact learned behaviors. This is called imprinting.

    - Example: geese have no innate understanding of "being" a goose". They will respond with the first object they encounter during a critical time period. If a human is introduced at a critical time period, the goose will prefer humans and may even identify enough with humans to attempt mating with them.

  - **Habituation** causes a lack of a response to repeated stimuli and therefore prevents organisms from wasting energy on irrelevant stimuli.

- In **classical conditioning** a stimulus that is totally arbitrary is associated with a reward or punishment.

- In **operant conditioning** an organism learns by trial and error. Often rewards or punishments tend to cause organisms to learn to repeat or cease a particular behavior.

  - **Trial and Error** learning occurs when a reward or punishment is the end result of an attempt to succeed at a particular task.

- **Reasoning, or insight**, is the most complex form of learned behavior and is very complex in its design.

- *Forms of behavior determined by genes are subject to evolution by natural selection.*

  o If the behavior provides an adaptation that promotes reproductive success of an individual, it will impact the evolution of the species.

    - **Territoriality** occurs when an organism defends an area where resources are located.
    - Cooperative, social behaviors are important in the promotion of the species.

      - In **herds**, the group of organisms acts together, responding in the same way to a stimulus or acting by imitation.
      - **Dominance hierarchies** occur within groups, however, the individual within those groups tend to have specific roles. Those with more status achieve more mates, food, shelter and other necessities. This promotes the reproductive success of those individuals.
      - **Sexual selection** occurs when organisms choose mates in a nonrandom fashion. This competition drives the evolution of certain traits.

  o The behavior may also decrease the reproductive success of an individual, but promote the success of the species.

    - **Altruism** is an example of this behavior. Individuals may sacrifice themselves in order to promote the success of the species.

BIOGEOCHEMICAL CYCLES

- *Chemicals cycle through the biological and geological worlds with the energy for this movement being provided by the sun or heat from the center of the Earth.*

  o The water cycle drives the movement of water through precipitation, condensation and evaporation/transpiration.

    - **Evaporation** occurs as water changes from a liquid to a gas.
    - **Transpiration** is the evaporation of water from the surface of a leaf.
    - **Condensation** occurs as water changes from a gas to a liquid and droplets form

- **Precipitation** occurs after water condenses into droplets and falls to Earth in the form of rain, snow, sleet, or hail.

- *Respiration and photosynthesis are the main processes within the carbon cycle.*

  o The reactants of photosynthesis are sunlight, water and carbon dioxide. The major products of photosynthesis are sugars and oxygen.
  o Respiration is the reverse of photosynthesis, with the reactants being sugar and oxygen and the products being energy (ATP), water and carbon dioxide.
  o Carbon exists primarily as CO_2 in the atmosphere. It is converted into carbohydrates through photosynthesis.
  o Carbon is returned to the atmosphere through the following processes:

    - **Respiration:** As macromolecules are broken down, carbon dioxide is released.
    - **Decay:** Fungi and bacteria break down carbon in dead animals and plants and convert the carbon back into carbon dioxide or methane.
    - **Combustion:** The burning of fossil fuels releases carbon dioxide into the atmosphere, a major factor in the greenhouse effect. Many organisms are not able to use nitrogen in the gaseous form, so they require a symbiotic relationship in order to obtain nitrogen in a usable form. Dissolved carbon dioxide may also be released into the atmosphere as it changes to a gas at the surface of warm bodies of water.

  o Many organisms are not able to use nitrogen in the gaseous form, so they require a symbiotic relationship in order to obtain nitrogen in a usable form.

    - Nitrogen, in its gaseous form, is converted in the soil and water to ammonia and ammonium in the process of nitrogen fixation.

      • **Biological nitrogen fixation** is mediated on land by bacteria that live in the roots of legumes.
      • In water, nitrogen can be fixed by cyanobacteria.
      • The ammonia is then incorporated into amino acids, proteins, vitamins and nucleic acids.

MULTIPLE CHOICE QUESTIONS

For questions 1-4, use the following answer choices.

    (A) O_2
    (B) CO_2
    (C) glucose
    (D) rubisco
    (E) glyceraldeyde-3-phosphate

1. Product of cellular respiration and reactant of photosynthesis

2. Product of photosynthesis that is starting material for glycolysis

3. Fixes carbon dioxide to RuBP

4. Final electron acceptor of the electron transport chain in cellular respiration

For questions 5-8, use the following answer choices.

    (A) convergent evolution
    (B) divergent evolution
    (C) diversifying selection
    (D) stabilizing selection
    (E) directional selection

5. Sharks and dolphins having grey, streamlined bodies

6. Black peppered moths being eaten more easily than white because they are easier to see against the white birch trees

7. Wings of butterflies and wings of bats

8. Heterozygous condition for sickle cell being favored in African countries

For questions 9-10, use the following answer choices.

    (A) K species
    (B) r species
    (C) carrying capacity
    (D) logistic growth
    (E) exponential growth

9. Growth due to no resource limitations

10. Maximum population size that can be sustained in any given environment

For questions 11-15, use the following answer choices.

(A)   classical conditioning
(B)   operant conditioning
(C)   habituation
(D)   fixed action pattern
(E)   imprinting

11. Birds exposed to a full song in a critical period cause development of the same song.

12. Prairie dogs who live near hiking trails do not give alarm calls every time a person walks by.

13. Your dog learns to shake hands with you when you hold out your hand after being rewarded with treats.

14. A rat learns how to get through a maze when it is rewarded with a treat, but never obtains food if it makes a mistake.

15. A baby duck runs for cover when an object is thrown over its head, but stops running after many trials.

16. Which of the following processes includes the conversion of carbon dioxide to carbohydrates?

(A)   photosynthesis
(B)   respiration
(C)   precipitation
(D)   hydrolysis
(E)   catabolism

17. Higher plants absorb nitrogen from the soil in the form of

   (A) gaseous nitrogen
   (B) nitrites
   (C) nitrates
   (D) ammonia
   (E) amino acids

18. Which is the correct sequence of events as water goes from ocean to land in the water cycle?

   (A) Transpiration, precipitation, condensation, runoff
   (B) Evaporation, condensation, precipitation, runoff
   (C) Runoff, precipitation, evaporation, condensation
   (D) Precipitation, runoff, evaporation, transpiration
   (E) Evaporation, precipitation, runoff, transpiration

19. All of the following statements about dominance hierarchies are true, EXCEPT

   (A) dominance hierarchies determine the priority of access to food and shelter.
   (B) individuals high in the pecking order tend to have a neat appearance and are more confident.
   (C) dominance hierarchies reduce the breeding population.
   (D) dominance hierarchies tend to have a decreased expenditure of lives and energy.
   (E) once established, individuals within dominance hierarchies will display their superiority through battle.

20. Which of the following examples describes a commensal relationship?

   (A) *E.coli* that inhabits the digestive tract of humans absorbs nutrients from its host, enters the blood stream and causes disease.
   (B) Rhizobium form nodules on the roots of a legume, releasing fixed nitrogen into the plant's cytoplasm.
   (C) In the tropical rain forest, epiphytes grow on the top of large trees, receiving more sunlight and getting minerals from the leaves that fall from the trees, while not harming nor helping the tree.
   (D) Mistletoe lives on oak and cedar trees, and while it can photosynthesize, gains minerals and water by invading the tissues of the tree and causing damage.
   (E) Protozoa live in the gut of the termite, digesting food and gaining protection from the termite.

FREE RESPONSE QUESTIONS

1. Human beings are characteristic K species. Yet, the growth of the human population is exponential and not logistic.

   a. **Identify** common features of K species?
   b. **Explain** why K species generally grow logistically?
   c. What types of organisms generally grow exponentially?
   d. **Discuss** what has allowed human beings population to continue to grow eponentially.

2. Competition is the interaction between two organisms for food, water, light, habitat, or mates. Direct competition between two species rarely continues for long periods of time.

   a. **Explain** three of the possible results from direct competition between two species.
   b. **Explain** how the presence of three of the following isolating mechanisms would contribute to the speciation of the organisms:

      i. Geographic barriers
      ii. Ecological isolation
      iii. Behavioral isolation
      iv. Polyploidy

Theme Eight

Science, Technology, and Society

BIOTECHNOLOGY

- *In the arena of biotechnology, new understandings in the intricate workings of DNA has led to an array of applications in medicine, industry, research, and environmental improvements..*
  - o **Genetic engineering** involves transferring DNA from one organism to another organism.
  - o **Transgenic organisms** are produced when DNA is transferred from one species to another. Transgenic organisms are important in medicine and in agriculture.
  - o The gene of interest is transferred through a vector into the host organism. Common vectors include transformed plasmids, transformed viruses, microinjection, or gene guns.
  - o **Restriction enzymes** are used to cut DNA samples from a desired source and place them into a new source (i.e. plasmid or virus). The complementary sticky ends of the DNA fragments allow the new genes to connect to the DNA of the vector.
  - o Transgenic pigs and cattle have been engineered with the genes for human growth factor, clotting factors, and human insulin. Then, these transgenic organisms produce these proteins to treat humans lacking these proteins.
  - o Transgenic plants can be resistant to rotting, freezing, or insect pests.
  - o Bacteria have been engineered to digest oil spills and other environmental pollutants through **bioremediation**.
  - o Plants can be used to extract toxins from the soil in **phytoremediation**. Then, the plant tissue can be removed from an ecosytem.
  - o **Electrophoresis** is a process where DNA fragments are separated by size to view bands of various sizes. In this process, restriction enzymes digest DNA into small fragments. The DNA samples are loaded into a gel medium and voltage is applied. The DNA fragments migrate to the positively charged pole of the electrophoresis apparatus due to the negative charge of the DNA fragments. Smaller molecules can move through the gel medium more rapidly and move further down the gel. Large fragments stay close to the loading wells.
  - o Electrophoretic bands can be analyzed to identify individuals, to compare genetic similarities of organisms, to discern guilty suspects in a criminal investigation, and to determine paternity. Comparing bands of different sizes is called **Restriction Fragment Length Polymorphism**.
  - o **Hybridization** is a process that uses radioactive or fluorescent nucleotide fragments to bind to a denatured DNA sequence in one fragment. This process can locate the fragment on a gel or nitrocellulose paper that has the desired gene or DNA sequence.

121

o **Polymerase Chain Reaction** (PCR) allows a small DNA sample to be amplified. A DNA sample is heated so that the double strand denatures and separates. Primers, DNA polymerase, and nucleotides are available to bind to the single DNA strands as the sample cools. While cooling, each original DNA strand is copied producing two DNA double strands. Then, the process is repeated over and over again to make many copies of the starting DNA.

o **DNA sequencing** is done to determine the sequence of DNA nucleotides in a DNA fragment. The original double helix is denatured. Then, primers, DNA polymerase, nucleotides, and dideoxynucleotides are needed. **Dideoxynucleotides** are nucleotides that have been altered to act as chain terminators. The 3' end of these nucleotides is altered so that another new nucleotide cannot be added to that end and when these dideoxynucleotides are present in a growing DNA strand, they stop the replication process. Four sample tubes are used: one for each type of dideoxynucleotide (ddA, ddT, ddG, and ddC). To each of these tubes, the other materials are added. Then, the single strands of DNA are replicated. Fragments of multiple lengths are produced in each tube due to the dideoxynucleotides. Then, electrophoresis is used to separate the strands by size and the sequence of nucleotides in the original DNA sample can be interpreted.

o **Microarrays** or gene chips are silicon chips that have many wells. In each well are strands of cDNA probes. cDNA is produced by using reverse transciptase to copy mRNA into DNA that is lacking intron regions. When a sample of tissue is placed on the chip, the mRNA from the sample binds to the complementary cDNA. Then, the level of binding and fluorescence of wells can determine the level of activity of a gene.

o **Gene therapy** is a process where functional genes are transferred to individuals through a vector (usually a virus). Individuals who have disorders like hemophilia or diabetes that are due to inactive genes can be treated by this process. The active gene will make the proteins that are lacking in the individual's cells.

o **RNA interference** is a process that uses siRNA to bind to mRNA that is produced from overactive genes. Disorders like cancer and Huntington's that are due to overactive genes could be treated by this process. The siRNA binds to the mRNA to block translation of the mRNA into proteins. Then, it signals the RISC complex to digest the mRNA.

o **Stem cells** are cells that have the potential to turn into a variety of cell types. These cells have the potential to repair damaged cells due to spinal cord injuries, neurological disorders like Parkinson's, strokes, or heart attacks. There are three classes of stem cells: totipotent, pluripotent, and multipotent. **Totipotent stem cells** come from embryos before the 8 cell stage. These cells can form any type of tissue and be used to clone new organisms. **Pluripotent stem cells** are extracted from embryos between the 8 cell stage and the formation of a blastula. These cells are completely undifferentiated. They can be used to form any type of tissue. However, they do not have the potential to turn into new organisms. **Multipotent stem cells** are extracted from embryos following gastrulation, umbilical cord blood, or adult bone marrow. These stem cells are partially differentiated and can form a limited number of types of new cells.

ENVIRONMENTAL CONSERVATION AND THE BIOSPHERE

- *Society's introduction of exotic species to an ecosystem has long lasting effects on other organisms and the environment.*
  - o An **exotic species** is any species not native to a particular environment that has the potential to cause harm to the organisms within the current ecosystems.
- *Sustaining genetic diversity is a critical factor in promoting the welfare of our society, as well as that of other organisms.*
  - o Threatened species may provide food, crops, pharmaceuticals and textiles for our society.
  - o The extinction of a particular species will cause us to lose any genetic potential from that species.
  - o We are dependent on other species and the extinction of those species may cause harm to our livelihood.
- *Restoring degraded areas should be a priority in our conservation efforts.*
  - o **Bioremediation** helps to improve the condition of polluted ecosystems by the use of living organisms, including bacteria and fungi.
  - o The determination of what nutrients have been removed from an ecosystem is critical for restoration.
  - o Restoration not only involves the reintroduction of species, nutrients, or other abiotic factors, but also an understanding of the roles of the organisms within the ecosystem and the interaction between species.
- *By placing societal needs in front of conservation and preservation, we are destroying the biosphere.*
  - o The accumulation of carbon dioxide and other greenhouse gases in the atmosphere are causing the **greenhouse effect**, a warming of global temperatures.
  - o Ozone absorbs UV radiation, reducing the levels of UV radiation reaching the earth's surface and causing mutation in the DNA of organisms. As we use chemicals (most notably CFC's) that break down the **ozone layer**, we reduce the ability of the ozone layer to prevent UV radiation from reaching the earth.

o **Deforestation** causes erosion, flooding, and changes in weather patterns, as well as a depletion of nutrients within the soil.
o **Biological magnification** occurs as toxins become increasingly concentrated in organisms higher up food chains. As the complexity of the food web increases, so does the intensity of the toxin.
o **Eutrophication** occurs as nutrient enrichment increases in lakes and subsequently increases the biomass at an accelerated rate.
  ■ Increased nutrients cause algal blooms.
  ■ Increased algae levels use up the nutrients, resulting in death of the algae and large quantities of organic matter.
  ■ Large quantities of decaying organic matter result in bacteria using up O_2
  ■ Loss of O_2 can kill fish and other organisms.

MULTIPLE CHOICE QUESTIONS

1. Restriction enzymes

   (A) denature DNA double helices
   (B) help DNA polymerase bind to single DNA strands
   (C) break DNA samples into small fragments
   (D) digest DNA into single nucleotides
   (E) encourage binding of radiaoactive probes to DNA fragments

2. Common vectors to transfer DNA from one species to another include all of the following EXCEPT

   (A) recombinant plasmid
   (B) virus
   (C) microinjection
   (D) gene gun
   (E) cDNA probe

3. RNA interference can be used to treat overactive genes by

   (A) preventing DNA replication
   (B) blocking translation and causing mRNA degradation
   (C) interfering with RNA polymerase binding to gene promoters
   (D) removing mRNA end caps
   (E) splicing the overactive gene sequence from the genome

For questions 4-7, select one of the following answer choices;

   (A) RNA interference
   (B) gene therapy
   (C) stem cell therapy
   (D) bioremediation
   (E) phytoremediation

4. A process to introduce functional genes when genes are inactive

5. Engineered microbes digest oil spills

6. Plants absorb chemicals and toxins from the soil

7. Introduce new cells to replace cells damaged by spinal cord injury

For questions 8-10, select one of the following answer choices:

   (A)   PCR
   (B)   DNA sequencing
   (C)   electrophoresis
   (D)   hybridization
   (E)   microarray

8. Process to amplify small DNA sample.

9. Uses cDNA probes to bind to mRNA.

10. Uses dideoxynucleotides to terminate replication.

11. Which of the following organisms are used in bioremediation techniques?

     I.   bacteria
    II.   fungi
  III.   protists

   (A)   I only.
   (B)   II only.
   (C)   III only.
   (D)   I and II.
   (E)   I and III.

For questions 12-16, select one of the following answer choices:

   (A)   eutrophication
   (B)   greenhouse effect
   (C)   ozone depletion
   (D)   biological magnification
   (E)   deforestation

12. Enhanced by the process of nutrient enrichment in lakes and streams.

13. The gradual warming of the biosphere due to an increase in carbon dioxide.

14. An accumulation of pesticides as the complexity of the food web increases.

15. The gradual increase of CFCs causes this to occur.

16. Results in increased penetration of UV radiation to the Earth's surface.

Use the following pyramid of energy to answer Questions 17-20.

(A)   hawk
(B)   snake
(C)   shrew
(D)   grasshopper
(E)   marsh grass

17. The trophic level where biological magnification of pesticides is the greatest.

18. A primary producer.

19. A primary consumer.

20. An autotroph.

FREE RESPONSE QUESTIONS

1. Denaturation of DNA involves heating the DNA double helix so that the two strands unwind. It is an important step in many biotechnological processes. For each of the following processes: PCR, DNA sequencing, electrophoresis, and hybridization.

   a. **Identify** whether the process requires denaturation and why it does or does not need denaturation for the process to be done

   b. Select two of the given four processes and **describe** how the process is done

   c. For the same two processes, **provide** and **discuss** current applications of that technology.

2. By placing societal needs in front of conservation and preservation, we risk destroying the biosphere. One of the ways in which we are doing this is through biological magnification.

   a. **Design** a pyramid of biomass for a typical freshwater ecosystem.

   b. **Discuss** how biological magnification of the pesticide DDT will impact this ecosystem.

Sample Examinations

GENERAL GUIDELINES

In this section, there are three complete exams that simulate the AP biology exam given by the College Board in May. For each exam, there is a 100-question multiple choice section and four-question free response section. These sections contain the same number and styles of questions that will appear on the official AP exam.

When taking the practice exams, allow 80 minutes to complete the 100 multiple choice questions. When the multiple choice section is scored, there is a penalty for incorrect answers. So, if no answer choices can be eliminated, the question should be skipped. If one or more answer choices can be eliminated, it is best to guess from the remaining choices. In scoring, ¼ of the number of incorrect responses is subtracted from the number of correct choices. Unanswered questions do not yield points or receive deductions.

Allow 10 minutes to read the free response questions and plan your writing without beginning the official writing process. You will be assigned 4 free response questions that must be answered. There isn't any question selection on the AP biology exam. This time period is designed to organize thoughts so that the time allotted for the essays can be used to write the essays.

Allow 90 minutes to write the four free response questions. Write in complete sentences and use paragraph format. Outlines and sketches are not awarded points.

Sample Examination I

Directions: Each of the following questions or statements is followed by five possible answers or sentence completions. Select the best answer or completion for each question or statement.

1. Translocation of xylem sap is most affected by

    (A) root pressure
    (B) source-sink dynamics
    (C) osmotic potential in sieve tube members
    (D) transpirational pull
    (E) cell wall component of tracheids

2. Which structure is most responsible for dramatically increasing the surface area available for respiration in mammals?

    (A) nephron
    (B) capillary
    (C) alveolus
    (D) gill
    (E) villus

3. Which organism uses Malphigian tubules for excretion?

    (A) earthworm
    (B) planarian
    (C) cricket
    (D) elephant
    (E) swordfish

4. Organisms with radial symmetry as juveniles and as adults belong to the Phylum

    (A) Porifera
    (B) Cnidaria
    (C) Platyhelminthes
    (D) Mollusca
    (E) Echinodermata

131

5. A form of asexual reproduction that results in dissimilar haploid offspring from the development of adults from unfertilized eggs is

    (A) parthenogenesis
    (B) apomyxis
    (C) regeneration
    (D) binary fission
    (E) budding

6. Which organism has a cell wall of chitin?

    (A) jellyfish
    (B) fern
    (C) mushroom
    (D) amoeba
    (E) bacteria

7. All of the following are mechanisms to control protein formation in eukaryotes EXCEPT

    (A) operons
    (B) histone acetylation
    (C) DNA methylation
    (D) alternative splicing
    (E) DNA compaction

8. Dave suffers from red-green colorblindness and Mary is a carrier for red-green color-blindness,but does not express the trait. Red-green colorblindness is the result of a recessive X-linked allele. What percentage of their sons might be colorblind?

    (A) 0%
    (B) 25%
    (C) 50%
    (D) 75%
    (E) 100%

9. Which scientist team is **incorrectly** paired with its contribution to understanding the structure and function of DNA and protein

   (A) Beadle and Tatum—each gene codes for one polypeptide
   (B) Watson and Crick—Four levels of protein structure
   (C) Hershey and Chase—DNA is the transforming agent with bacteriphages
   (D) Meselson and Stahl—DNA replication is semiconservative
   (E) Avery, McCarty, and MacLeod—DNA is transforming agent with bacteria

10. The oxygen independent stage of cellular respiration that occurs outside of the mitochondrion is the stage of

    (A) glycolysis
    (B) pyruvate oxidation
    (C) Krebs cycle
    (D) electron transport
    (E) chemiosmosis

11. All of the following are methods for genome alteration to occur in bacteria EXCEPT

    (A) bacteria can uptake foreign DNA from the environment through transformation
    (B) bacteria can access new bacterial DNA through a virus intermediate through transduction
    (C) mutation can occur
    (D) bacteria can splice genes through the lactose operon
    (E) bacteria can access DNA from other bacteria through conjugation

12. A cytoplasmic bridge called a pilus is necessary for the process of

    (A) mutation
    (B) transduction
    (C) conjugation
    (D) transformation
    (E) operon regulation

13. Which one of the following is the best example of Batesian mimicry?

    (A) A palatable viceroy butterfly is colored the same as a toxin-filled monarch butterfly
    (B) Two species of stinging wasps have the same coloring pattern
    (C) Blue-jean poison dart frogs have brightly-colored blue and red skin
    (D) Two palatable species of ants undergo the same colony building practices
    (E) A peppered moth blends in with the bark of a birch tree

14. Hydrogen bonds form readily between

    (A) two nonpolar molecules
    (B) two polar molecules
    (C) two ionic compounds
    (D) one polar molecule and one nonpolar molecule
    (E) two steroid molecules

15. ATP serves as the primary source of cellular energy because it

    (A) has many carbons and nitrogens
    (B) has triple covalent bonds between the nitrogens
    (C) has three bound phosphate groups
    (D) has a relatively small molecular size
    (E) is easy for cells to build and hard to break

16. During mitosis, the sister chromatids are pulled toward opposite centrosomes during

    (A) prophase
    (B) prometaphase
    (C) metaphase
    (D) anaphase
    (E) telophase

17. Which one of the following statements is true of a human somatic cell?

    (A) Each one is genetically unique.
    (B) Each one contains only maternal chromosomes.
    (C) Each one is diploid.
    (D) Each one has the same basic cellular function.
    (E) It fuses with the opposing gamete in fertilization.

18. A nondisjunction happens when

    (A) two chromosomes fail to separate completely during meiosis
    (B) a piece of one chromosome breaks off and attaches to a nonhomologous
         chromosome
    (C) a piece of one chromosome is lost
    (D) the gene order on a chromosome is inverted
    (E) a single base pair is deleted from a gene sequence

19. A major mechanism that gives rise to sympatric speciation in plants is

    (A) apomyxis
    (B) apoptosis
    (C) parthenogenesis
    (D) polyploidy
    (E) phytoremediation

20. The enzyme helicase

    (A) builds a new DNA strand along its template
    (B) connects Okazaki fragments together
    (C) removes introns from mRNA
    (D) unwinds the original DNA double helix to begin replication
    (E) builds a mRNA strand along the DNA template

21. The nephron is a tubular structure of the kidney. As filtrate flows through the nephron tubule, it flows in a specific direction. The correct passage of filtrate through the nephron tubule is

    (A) Bowman's capsule, loop of Henle, proximal convoluted tubule, distal convoluted tubule
    (B) loop of Henle, proximal convoluted tubule, Bowman's capsule, distal convoluted tubule
    (C) Bowman's capsule, distal convoluted tubule, loop of Henle, proximal convoluted tubule
    (D) Bowman's capsule, proximal convoluted tubule, loop of Henle, distal convoluted tubule
    (E) proximal convoluted tubule, distal convoluted tubule, loop of Henle, Bowman's capsule

22. When birds eliminate nitrogenous waste, they eliminate that waste in a less soluble form than humans. Birds eliminate nitrogenous waste in the form of

    (A) uric acid
    (B) ammonia
    (C) nitrate
    (D) urea
    (E) nitrous oxide

23. Which of the following aquatic zones is incorrectly paired with it's description?

    (A) Intertidal zone—shallow area where the water meets land
    (B) Littoral zone—area of open water
    (C) Profundal zone—deep aphotic region
    (D) Neritic zone—shallow area above the continental shelf
    (E) Abyssal zone—benthic region that does not have any light

24. In most nutrient cycles, there is an atmospheric component of the cycle. Which of the following nutrients does not have an atmospheric component to its biogeochemical cycle?

    (A) oxygen
    (B) phosphorous
    (C) nitrogen
    (D) carbon
    (E) water

25. Blood has living cells (erythrocytes and leukocytes) embedded in serum (nonliving matrix). This type of tissue is

    (A) epithelial
    (B) muscular
    (C) skeletal
    (D) nervous
    (E) connective

26. The region of the small intestine where the most chemical digestion occurs is the

    (A) jejunum
    (B) ileum
    (C) cecum
    (D) colon
    (E) duodenum

27. Clams are suspension feeders. They would most likely obtain food particles by

    (A) trapping large prey
    (B) attaching to a food substrate to digest it
    (C) penetrating a host to suck blood
    (D) sifting small particles out of the water
    (E) absorbing decaying materials

28. The least specialized plant cells are responsible for most of the metabolic activity in plants. These fairly unspecialized cells with a thin cell wall are

    (A) sieve tube elements
    (B) tracheids
    (C) collenchyma cells
    (D) sclerenchyma cells
    (E) parenchyma cells

29. Angiosperms can be classified as monocots or dicots. All of the following are common characteristics of monocot plants EXCEPT

    (A) parallel venation
    (B) fibrous roots
    (C) one cotyledon
    (D) scattered vascular bundles in the shoot
    (E) floral parts in multiples of four or five

30. There is a high degree of diversity within the class Mammalia. What trait distinguishes monotremes from the other mammal groups?

    (A) ability to fly
    (B) pouches to further development
    (C) mammary glands
    (D) lay eggs
    (E) feathers in place of hair

31. During the process of development, it is essential for some cells to systematically shut down and commit cell suicide. This cell death allows for the removal of cells that are unnecessary to the organism as it grows. This programmed cell death is called

    (A) mutagenesis
    (B) apomyxis
    (C) parthenogenesis
    (D) apoptosis
    (E) conjugation

32. During translation, when a stop codon is read on the mRNA strand at the ribosome,

    (A) a concluding glycine is attached to the growing polypeptide
    (B) a repressor attached to the ribosome that inhibits the movement of the ribosome along the mRNA
    (C) the enzyme helicase binds to terminate the polypeptide
    (D) a release factor enters the A site of the ribosome and stimulates the disassembly of the translation complex
    (E) the mRNA is digested by a protease complex

33. When materials move across a cell membrane through facilitated diffusion,

    (A) the energy of ATP is needed
    (B) the molecules move from an area of low concentration to an area of high concentration
    (C) an integral protein is needed
    (D) the molecules are often linked to the passive transport of glucose molecules
    (E) the molecular movement is driven by a sodium-potassium pump

34. Alfred Hershey and Martha Chase were instrumental in building an understanding of DNA. They worked with bacteriophages to demonstrate that

    (A) DNA was the genetic material
    (B) DNA has a double helix structure
    (C) each gene codes for one protein
    (D) transposable elements exist
    (E) DNA replicates semi-conservatively

35. The semi-conservative replication of DNA means that

    (A) one double helix is maintained as a double helix and copied in its intact form
    (B) a double helix is separated and each single strand is copied to form two new double helices
    (C) spliceosomes will digest single strand pieces of DNA
    (D) the DNA double helix is broken into tiny fragments that are copied and then rebuilt to form two new double helices
    (E) the process of DNA replication is very different in different animal species.

36. The Human Genome Project has recently been completed. The primary goal of this endeavor was to

    (A) make a gene library of all of the plasmids that have been recombined with human genes
    (B) collate a database that files electrophoretic gel data for each human in a specific country
    (C) completely sequence all of the genes in a human
    (D) collect data about all of the human genes that appear in the same form in other organisms
    (E) identify all of the transposable elements that occur in mammals

37. Monosaccharides generally contain

    (A) only carbons and hydrogen
    (B) an equal number of carbons and oxygens
    (C) a carbon and nitrogen ring structure
    (D) an amino group and a carboxyl group
    (E) a carbon triple covalently bonded to a nitrogen

38. Sickle cell anemia is the result of single base pair substitution that alters one amino acid in the resulting protein. A change in the amino acid sequence represents a change in the _____ structure of the protein.

    (A) primary
    (B) secondary
    (C) tertiary
    (D) quaternary
    (E) quintary

39. Which one of the following events occurs during interphase of the human cell cycle?

    (A) Sister chromatids separate and move toward opposite poles
    (B) Homologous chromosomes synapse and cross-over
    (C) The nuclear envelope fragments
    (D) The DNA is replicated
    (E) The spindle fibers attach to the kinetochores of the chromatids

40. What percent of the children will have type O blood when a man with type O blood marries a woman with type AB blood?

    (A) 0%
    (B) 25%
    (C) 50%
    (D) 75%
    (E) 100%

41. If two individuals who were both heterozygous for three traits (A, B, and C) were crossed, what would be the probability of the following genotype in the first offspring: AAbbCc?

    (A) 1 out of 8
    (B) 1 out of 16
    (C) 1 out of 32
    (D) 1 out of 64
    (E) 1 out of 128

42. Open sinuses and hemolymph as the transport fluid are characteristics of a(n)

    (A) two circuit closed circulatory system
    (B) human circulatory system
    (C) open circulatory system
    (D) prokaryotic cell
    (E) plant vascular system

43. In mammals, the central nervous system directly controls the contraction of some muscle cells through the release of neurotransmitters. Autorhythmic control is common in other muscle cells. Autorhythmic control of contraction is common in

    i. smooth muscle cells
    ii. cardiac muscle cells
    iii. skeletal muscle cells

    (A) i only
    (B) ii only
    (C) i and iii only
    (D) ii and iii only
    (E) i, ii, and iii

44. All of the following are components of the nonspecific immune response EXCEPT

    (A) helper T cells
    (B) skin
    (C) complement proteins
    (D) phagocytic cells
    (E) mucous linings

45. The inflammatory response is caused by a chemical signal from basophils and mast cells. This signal causes inflammation as arterioles dilate and venules constrict. This inflammatory signal is

    (A) cytokine
    (B) lysozyme
    (C) performin
    (D) interleukin 1
    (E) histamine

46. All of the following are characteristics of organisms that tend to experience exponential growth except for

    (A) long period of parental care
    (B) many offspring at each reproductive episode
    (C) short lifespan
    (D) short generation time
    (E) poor competitors

47. In peas, tall plants (T) are dominant to short (t) and Purple flowers (P) are dominant to white (p). If two plants that are heterozygous for both height and flower color mate, what frequency of the offspring should be short plants with purple flowers?

    (A) 1/4
    (B) 1/16
    (C) 3/16
    (D) 9/16
    (E) 11/16

48. The endosymbiotic theory describes

   (A) the evolution of eukaryotic cells from prokaryotic ancestors
   (B) the evolution of the parasitic lifestyle in animals
   (C) the coevolution of termites and the gut protist that helps with the digestion of lignin
   (D) the divergence into separate kingdoms within the Domain Eukarya
   (E) the evolution of the first prokaryotic cells

49. Which of the following is an example of the bottleneck effect?

   (A) a small group of frogs moves to open pond in another ecosystem to start a new population
   (B) a severe flood kills 98% of a successful cricket population leaving only a very small percentage to rebuild the population
   (C) sea gulls easily move between two different island populations
   (D) female peacocks only select males with the most impressive plumes
   (E) white peppered moths are favored and black moths are selected against

50. An organelle that is the site of protein synthesis in both eukaryotes and prokaryotes is the

   (A) nucleus
   (B) golgi apparatus
   (C) rough endoplasmic reticulum
   (D) smooth endoplasmic reticulum
   (E) ribosome

Directions: There are five lettered headings for each group of questions, followed by a list of phrases or sentences. For each phrase or sentence, select the best answer. Answers may be used once, more than once, or not at all in each group.

For questions 51-55, select a biome from the following answer choices.

   (A) tundra
   (B) taiga
   (C) sesert
   (D) tropical rainforest
   (E) temperate grassland

51. Permafrost remains under the soil

52. Hot days and cold nights

53. CAM plants are the dominant plant type

54. Highest degree of species diversity

55. Pine trees are the dominant plant type

For questions 56-60, select an answer from the following choices. Select answers based upon a standard terrestrial community.

    (A) producer
    (B) herbivore
    (C) decomposer
    (D) omnivore
    (E) top carnivore

56. The highest biomass

57. The smallest abundance

58. Feeds on organisms in every trophic level

59. The animal group with the highest number of organisms

60. Organisms that feed on the producers and the herbivores

For questions 61-65, select an answer choice from the following choices.

    (A) facilitated diffusion
    (B) active transport
    (C) osmosis
    (D) endocytosis
    (E) exocytosis

61. Movement of water molecules from high water concentration to low across a semipermeable membrane

62. Movement of a material from low concentration to high through an integral protein

63. Bulk movement of particles into a cell

64. Bulk movement of particles out of a cell

65. Movement of ions from high concentration to low through a protein

For questions 66-70, select an answer choice from the following choices.

    (A) ribosome
    (B) chloroplast
    (C) lysosome
    (D) mitochondrian
    (E) cytoskeleton

66. Site of cellular respiration

67. Site of translation

68. Site of photosynthesis

69. Contains hydrolytic enzymes

70. Composed of microtubules and microfilaments

For questions 71-75, select an answer choice from the following choices.

    (A) glucagon
    (B) insulin
    (C) calcitonin
    (D) parathyroid hormone
    (E) oxytocin

71. Produced by the thyroid gland

72. Helps to reduce blood sugar levels

73. Helps to increase blood sugar levels

74. Helps to reduce blood calcium levels

75. Helps to increase blood calcium levels

Analyze the table, graph, figure or diagram. Then, use that figure to answer the questions that follow.

Dialysis bags with various concentrations of sucrose were placed into a beaker with a beaker of sucrose of unknown concentration. The results are recorded below.

Bag Number	Sucrose Concentration in the Bag (%)	Initial Weight of Dialysis Bag (g)	Final Weight of Dialysis Bag (g)
1	0	12.5	4.5
2	5	12.5	8.9
3	10	12.5	11.4
4	15	12.5	12.3
5	20	12.5	13.8
6	25	12.5	16.2

76. Bag number 1 decreased in mass as

    (A) sucrose moved out of the bag
    (B) sucrose moved into the bag
    (C) water moved out of the bag
    (D) water moved into the bag
    (E) balancing salts moved out of the bag

77. The concentration of the sucrose in the beaker is

    (A) 4.5%
    (B) 8.9 %
    (C) 11.4%
    (D) 15.2%
    (E) 26.2%

78. At the beginning of the experiment, which word best describes a comparison of the solution in bag #6 to the solution in the beaker?

    (A) isotonic
    (B) hypotonic
    (C) hypertonic
    (D) amphipathic
    (E) homeotic

A DNA sample was digested by three different restriction enzymes: EcoRI, BamHI, and HindIII, and pushed through an electrophoretic gel. The base pair sizes of the fragments are presented in the table.

Lane 1: EcoRI	Lane 2: BamHI	Lane 3: HindIII
22,245 bp	19,456	18,458
16,345 bp	18,842	14,396
9,460 bp	11,285	9,217
6,635 bp	9,290	6,258
2,228 bp	8,343	4,348

79. In lane 1, which fragment would migrate the farthest through the gel?

    (A) 22,245 bp
    (B) 16,345 bp
    (C) 9,460 bp
    (D) 6,635 bp
    (E) 2,228 bp

80. Which fragment will remain closest to the starting wells?

    (A) 22,245 bp
    (B) 18,842 bp
    (C) 8,343 bp
    (D) 4,348 bp
    (E) 18, 458 bp

81. The restriction enzymes used to digest the DNA in this exercise were probably harvested from

    (A) humans
    (B) ferns
    (C) mushrooms
    (D) bacteria
    (E) viruses

Use the genetic code table to answer the following questions.

					Second mRNA base		
		U	C	A	G		
First	U	Phe	Ser	Tyr	Cys	U	**Third**
		Phe	Ser	Tyr	Cys	C	
		Leu	Ser	Stop	Stop	A	
		Leu	Ser	Stop	Trp	G	
mRNA	C	Leu	Pro	His	Arg	U	**mRNA**
		Leu	Pro	His	Arg	C	
		Leu	Pro	Gln	Arg	A	
		Leu	Pro	Gln	Arg	G	
Base	A	Ile	Thr	Asn	Ser	U	**Base**
		Ile	Thr	Asn	Ser	C	
		Ile	Thr	Lys	Arg	A	
		Met Start	Thr	Lys	Arg	G	
	G	Val	Ala	Asp	Gly	U	
		Val	Ala	Asp	Gly	C	
		Val	Ala	Glu	Gly	A	
		Val	Ala	Glu	Gly	G	

82. If DNA samples were amplified before digestion with restriction enzyme to increase the amount of DNA available for analysis, the DNA would be amplified through

    (A) RFLP
    (B) VNTR
    (C) DNA sequencing
    (D) RNA interference
    (E) PCR

83. What mRNA strand would form from a DNA template strand of TACGCACCCACG?

    (A) TACGCACCCACG
    (B) UACGCACCCACG
    (C) ATGCGTGGGTGC
    (D) AUGCGUGGGUGC
    (E) UTGCGTGGGTGC

84. What amino acid sequence would form from a DNA template strand of TACGCACCCACG?

    (A) Val-Asn-Arg-Ser
    (B) Met-Arg-Gly-Cys
    (C) Val-Arg-Ser-Cys
    (D) Met-Gly-Ser-Asn
    (E) Ile-Leu-Val-Ala

85. The anticodon on the tRNA molecule that will bring Serine to a ribosome could be

    (A) AUC
    (B) AGC
    (C) UCC
    (D) UUA
    (E) CCG

86. What DNA template would yield Gly Ala Arg Pro Val?

    (A) GGCGCCCGACCAGUA
    (B) CCGCGGGCTGGTCATATT
    (C) GGCGCCCGACCAGTA
    (D) CGCACCGGCTTATTC
    (E) CCGCGGTTAGGATTT

87. Which statement is true about the genetic code?

    (A) The genetic code is highly variable and differs greatly between species.
    (B) The genetic code is conserved within major kingdgoms, but varies greatly between kingdoms.
    (C) The genetic code is conserved with each domain, but varies greatly between the three domains.
    (D) The genetic code is highly conserved and very similar in all organisms.
    (E) This genetic code table is specific to human beings.

Answer the following questions about this sample grassland foodweb.

Grass → Cricket → Frog → Snake → Owl

88. The amount of available energy is the highest for the

    (A) grass
    (B) cricket
    (C) frog
    (D) snake
    (E) owl

89. The primary producer is the

    (A) grass
    (B) cricket
    (C) frog
    (D) snake
    (E) owl

90. The primary consumer is the

    (A) grass
    (B) cricket
    (C) frog
    (D) snake
    (E) owl

91. The amount of energy available to the owl population is about

    (A) 1/10 the energy of the grass
    (B) 1/100 the energy of the grass
    (C) 200 times the energy of the frog
    (D) 1/10 the energy of the snake
    (E) 100 times the energy of the snake

92. A food web differs from this simple food chain by all of the following EXCEPT

    (A) it contains producers
    (B) it contains decomposers
    (C) it shows organisms feed on more than one trophic level
    (D) it shows more than one species to fill each trophic level
    (E) it shows organisms can be omnivorous

Examine the following graph of the growth of a ground squirrel population in the Aroura State Park.

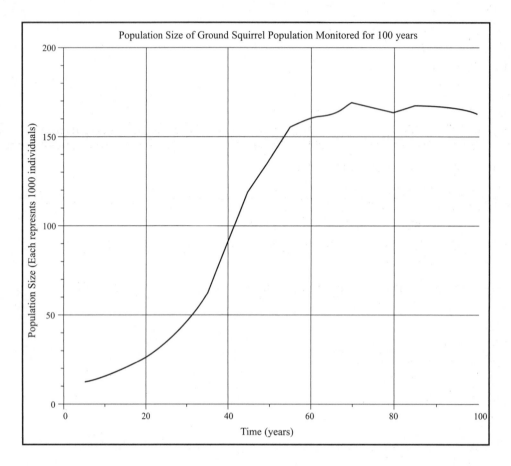

93. Which statement is true about this ground squirrel population during this 100 year period?

   (A) It continued to grow exponentially.
   (B) It began with logistic growth, but after about 50 years grew exponentially.
   (C) It experienced logistic growth.
   (D) It took 100 years for the population to begin growing exponentially.
   (E) It took 100 years for the population to begins showing signs of logistic growth.

94. The carrying capacity of this population is closest to

   (A) 50,000 individuals
   (B) 75,000 individuals
   (C) 105,000 individuals
   (D) 140,000 individuals
   (E) 180,000 individuals

95. How many years did it take for this population to reach its carrying capacity?

    (A) 10 years
    (B) 25 years
    (C) 40 years
    (D) 60 years
    (E) 100 years

96. Which one of the following traits probably does not apply to this population?

    (A) Strong ecological competitor
    (B) Parental care
    (C) Fairly long lifespan
    (D) Highly opportunistic
    (E) More than one reproductive episode

97. All of the following contribute to the increase in greenhouse gases EXCEPT

    (A) burning of fossil fuels
    (B) deforestation
    (C) cellular respiration
    (D) photosynthesis
    (E) automobile exhaust

98. The role that the greenhouse gases play in global warming is that they

    (A) reflect more of the sun's energy back into space
    (B) absorb water vapor and remove it from the air
    (C) absorb energy from rays that are reflected from earth's surface
    (D) breakdown the ozone molecules in the atmosphere
    (E) cause carbon pollution in the earth's water systems

99. Ozone molecules are important in the atmosphere because the ozone molecules

   (A) return reflected rays back to the earth
   (B) reduce the ultraviolet radiation that reaches the earth
   (C) replenish the oxygen gas needed by organism
   (D) convert atmospheric nitrogen to nitrates and ammonium
   (E) provide polar ice cap insulation that prevents melting

100. A major ozone depleting molecule is common in aerosols and refrigerants. This ozone damaging molecule is a(n)

   (A) carbon dioxide
   (B) nitrous oxide
   (C) chlorofluorocarbon
   (D) uric acid
   (E) ammonia

FREE-RESPONSE QUESTIONS

Directions: Answer all questions

You must answer the questions in essay form (no outlines). You may use diagrams and drawings, but they must be accompanied by discussion. Read each question thoroughly before you begin to write.

1. A well-defined excretory system is highly important for animals to rid their bodies of metabolic waste.

    a. **Describe** the structure and function of four evolutionary advancements of the mammalian excretory system.
    b. **Compare** the mammalian excretory system to that of an insect **OR** an earthworm.

2. Both animals and plants must be able to defend themselves from other organisms in their communities. Both plants and animals are at risk of being consumed as food by another organism.

    a. How do plants defend themselves against herbivory?
    b. How do animals defend themselves against predation?
    c. **Describe** evolutionary advancements of herbivores and predators to efficiently catch and utilize food sources.
    d. **Discuss** how invasive species disrupt predator/prey or herbivore/plant dynamics.

3. Enzymes are important biological molecules. They are one of the primary means of regulating chemical processes within cells.

    a. **Describe** how enzymes affect chemical reactions.
    b. **Describe** environmental factors that affect enzyme action.
    c. **Describe** how enzymes are important for the process of DNA replication.

4. A population of green slugs was measured at five different times. The body coloring in these slugs could be green striped (dominant) or green without stripes (recessive).

Measurement Times	Striped Slugs	Unstriped slugs
1	450	150
2	600	200
3	900	300
4	1000	500
5	1200	850

a.   Is this population experiencing Hardy-Weinberg equilibrium? **Explain** your answer.

b.   What could be happening in the population to yield the equilibrium or nonequilibrium that is being experienced?

c.   Is there evidence of microevolution in this population? **Explain** the evidence.

DATA CORRELATION: Sample Examination #1

After you have completed your practice exam, circle the numbers of the items you missed. Use this information to guide you in determining which topics may require a more in-depth review.

I. Molecules and Cells

Cellular Respiration	10
Molecules and Bonding	14,15,37,38
Membrane Transport	33,61,62,63,64,65,76,77,78
Organelle Structure and Function	50,66,67,68,69,70

II. Genetics and Evolution

Molecular Genetics (DNA and RNA)	7,9,11,12,18,20,31,32,34,35,83,84,85,86,87
Inheritance Patterns	8,40,41,47
Nuclear Division	16,39
Biotechnology	36,79,80,81,82
Evolution	19,48,49

III. Organisms and Populations

Plant Structure and Function	1,28,29
Animal Structure and Function	2,3,5,17,21,22,25,26,27,42,43,44,45,71,72,73,74,75
Classification	4,6,30
Ecology	13,23,24,46,51,52,53,54,55,56,57,58,59,60,88,89,90, 91,92,93,94,95,96,97,98,99,100

NO TESTING MATERIAL PRINTED ON THIS PAGE

Sample Examination II

Directions: Each of the following questions or statements is followed by five possible answers or sentence completions. Select the best answer or completion for each question or statement.

1. The plant hormone that is responsible for phototropism by stimulating cell elongation along the dark side of the plant is

    (A) auxin
    (B) cytokinin
    (C) gibberellin
    (D) abscisic acid
    (E) ethylene

2. Bundle sheath cells are an adaptation for photosynthesis of

    (A) C_3 plants
    (B) C_4 plants
    (C) CAM plants
    (D) C_3 and C_4 plants
    (E) C_4 and CAM plants

3. The most abundant protein on earth is an enzyme that binds cardon dioxide to ribulose bisphosphate in the Calvin Benson Cycle. This important plant enzyme is

    (A) rubisco
    (B) collagen
    (C) dehydrogenase
    (D) catalase
    (E) G3P

4. Hemophilia is a recessive sex-linked trait. If a man that does not have hemophilia marries a woman that does have hemophilia, what percentage of their sons will have hemophilia?

(A) 0%
(B) 25%
(C) 50%
(D) 75%
(E) 100%

5. Hemophilia is a recessive sex-linked trait. If a man that does not have hemophilia marries a woman that does have hemophilia, what percentage of their daughters will have hemophilia?

(A) 0%
(B) 25%
(C) 50%
(D) 75%
(E) 100%

6. Which one of the following is the best example of Batesian mimicry?

(A) A nonpoisonous king snake has the same colored stripes as a poisonous coral king
(B) Two poison dart frogs have the same bright green and red markings
(C) A walking stick blends in well with the branches of trees
(D) Two spider species have similar mating dance rituals
(E) Two bee species have similar yellow and black banding patterns

7. Which one of the following is the best example of learning by habituation?

(A) A dog barks when the doorbell rings and when the alarm sounds
(B) A dog wags his tail when the doorbell rings expecting to be pet by the new visitor
(C) A dog that was initially roused by the telephone ringing learns to ignore and not respond to the ringing
(D) A dog salivates every time that a new food source is presented to
(E) A dog salivates every time that food is removed from his bowl and from his visual area

8. The detritivores in a food web

   (A) generally are the primary producers
   (B) are the top predator of the web
   (C) are the organisms that feed on dead material from every level of the food web
   (D) are the herbivores that feed on the primary producers
   (E) have the most energy available to them of any trophic level

9. The endosymbiotic theory describes the evolution of

   (A) multicellular organisms from unicellular ancestors
   (B) the first autotrophs from heterotrophic ancestors
   (C) the first saprobes from herbivorous ancestors
   (D) glycolysis from early energy forming pathways
   (E) eukaryotic cells from prokaryotic ancestors

10. *C. elegans* is a microscopic roundworm that has become a model organism for research. All of the following are common traits of *C. elegans* that make it a model research organism EXCEPT

   (A) the genome has been successfully mapped
   (B) it is very small and reproductively prolific
   (C) it has a transparent body
   (D) the adult body contains a small number of cells and the developmental origin of each of those cells is clearly defned
   (E) its genetic material mutates at a very rapid rate when exposed to UV radiation.

11. Genomic imprinting happens when

   (A) patterns of alternative splicing are passed from parent cell to daughter cell
   (B) DNA methylation patterns are passed from parent cell to daughter cell
   (C) all regulatory operons are mechanically disassembled
   (D) mRNA binds to the large ribosomal subunit
   (E) DNA polymerase replaces mismatched nucleotides with correct ones

12. Movement through the mammalian digestive system is most achieved through

   (A) skeletal muscle contraction in the abdominal walls
   (B) asynchronous smooth muscle contraction in alimentary canal walls
   (C) enzymatic activity in the duodenum
   (D) osmotic potential in the cells of the organ walls
   (E) accessory organ activity

13. In mammals, increased surface area available for respiration is achieved by

    (A) having the lung tissue composed of tiny air pockets called alveoli
    (B) having small respiratory organs located throughout the body rather than one large respiratory center
    (C) capillaries converge into veins
    (D) cilia lining the respiratory passages
    (E) mucus lining all respiratory organs

14. Organisms with radial symmetry often

    (A) show high levels of cephalization
    (B) have keen sensory functioning
    (C) are sessile
    (D) have deuterostomic developmental patterns
    (E) lack any type of true body tissues

15. In the alternating life cycle found in plants, the two multicellular stages are

    (A) haploid sporophyte and diploid gametophyte
    (B) haploid gametophyte and diploid sprorophyte
    (C) haploid sporophyte and haploid gametophyte
    (D) diploid sporophyte and diploid gametophyte
    (E) male gamete and female gamete

16. Which of the following actions is one of the major contributors to the genetic uniqueness in the production of daughter cells by meiosis

    (A) alignment of homologous pairs along the metaphase plate
    (B) crossing-over between homologous chromosomes
    (C) fragmentation of the nuclear envelope
    (D) replication of the centromeres
    (E) attachment of the spindle fibers to the kinetochores

17. Which one of the following would describe an example of allopatric speciation?

    (A) Polyploidy creating a new species of strawberry
    (B) Apple maggots diverging into two species due to a food preference for apple or hawthorne trees that are distributed in the same area
    (C) Two diverging cicada populations due to the mating season of one group being the first week in March and the mating season of the other group is the last week in March
    (D) When a section of a large pond dries, it divides the pond into two distinct environments. After the separation the ancestral population of sunfish becomes distinctly different in the two ponds
    (E) Snails with spiral shells get selected more as mates than snails with smooth shells

18. When doing electrophoresis, DNA fragments are loaded near the cathode of the electrophoresis chamber because

    (A) DNA fragments carry a negative charge
    (B) DNA fragments have a high molecular weight
    (C) large DNA fragments move through the gel very quickly
    (D) the anode of the chamber has negative charge
    (E) DNA fragments are cations

19. Dideoxynucleotides are used to

    (A) separate the DNA double helix before PCR
    (B) separate the DNA double helix before DNA sequencing
    (C) act as chain terminators during DNA sequencing
    (D) act as chain terminators during PCR
    (E) be radioactive probes during Southern hybridization

20. All of the following features are common to both mammals and reptiles EXCEPT for

    (A) amniote egg
    (B) internal fertilization
    (C) keritin body covering
    (D) vertebrae
    (E) hair

21. The primary disadvantage of the two-chambered heart found in fish is that

    (A) there is substantial mixing of oxygen-rich and oxygen-deficient blood
    (B) there is a loss of pressure as blood moves away from the gills
    (C) there aren't any valves to regulate blood flow
    (D) blood flows directly from the heart to the veins
    (E) blood from the gills returns to the heart for a second circuit

22. Which one of the following is an example of primary growth in a plant?

    (A) Expansion of the cork cambium in an evergreen tree
    (B) Production of new xylem and phloem tissue
    (C) Cell division in the vascular cambium of an oak tree
    (D) Cellular division in the apical meristem of a tulip
    (E) Increase in the diameter of a tomato plant root

23. Even though the Krebs Cycle of cellular respiration does not require oxygen directly. It will stop happening when oxygen is absent. The best explanation for this cycle interruption is that

    (A) too many ATP molecules have accumulated from glycolysis
    (B) there is an excess of carbon dioxide molecules in the atmosphere
    (C) there is a lack of NAD+ molecules
    (D) there is an abundance of substrate level phosphorylation
    (E) the necessary enzymes are blocked by competitive inhibitors

24. Mycorrhizae are best described as a symbiotic relationship between a(n)

    (A) green alga and a bacterium
    (B) tapeworm and a human
    (C) plant and a fungus
    (D) plant and a bacterium
    (E) termite and gut protist

25. Root nodules found in legumes contain

    (A) photosynthetic bacteria
    (B) bacteria that fix atmospheric nitrogen
    (C) sugar reserves from excess photosynthesis
    (D) tissue that produces the major plant hormones
    (E) thousands of sperm containing pollen grains

26. During normal reproduction in mammals, fertilization of the ovum by a spermatid usually occurs in the

    (A) uterus
    (B) ovary
    (C) fallopian tube
    (D) vagina
    (E) vestibule

27. All of these events of digestion occur in the stomach of mammals EXCEPT

    (A) pepsin begins the digestion of proteins
    (B) gastric acid is released by the epithelial cells
    (C) acid chyme is neutralized by salts from the pancreas
    (D) volatile wall contractions causes churning
    (E) opening of the pyloric sphincter allows acid chyme to move to the duodenum

28. All of the following are major features of the members of phylum chordata EXCEPT for

    (A) notochord
    (B) dorsal hollow nerve chord
    (C) pharyngeal gill slits
    (D) radial symmetry
    (E) postanal tail

29. Organisms that are multicellular, heterotrophic saprobes with cell walls of chitin probably belong to

    (A) Domain Bacteria
    (B) Phylum Chlorophyta
    (C) Kingdom Plantae
    (D) Kingdom Fungi
    (E) Class Mammalia

30. Mitosis produces a variety of cell types in organisms. Mitosis does result in the production of

    (A) gametes in birds
    (B) spores in ferns
    (C) gametes in pine trees
    (D) prokaryotic daughter cells
    (E) zygotes in fish

31. Which organism would most likely have a predominantly haploid life cycle?

    (A) fern
    (B) mushroom
    (C) sea urchin
    (D) pinetree
    (E) earthworm

32. A chromosomal translocation happens when

    (A) two sister chromatids fail to separate during anaphase II of meiosis
    (B) a section of one chromosome is lost during meiosis
    (C) a section of one chromosome breaks off and attaches to a nonhomologous chromosome
    (D) the sequence of genes on one arm a chromosome are inverted
    (E) a point mutation occurs that causes a frameshift

33. Nondisjunction of chromosomes can result in genetic defects like Klinefelter's syndrome or Down's syndrome. A nondisjunction happens when

    (A) two sister chromatids exchange some genetic information
    (B) two sister chromatids or two homologous chromosomes fail to separate
    (C) one section of the long arm of a chromosome is lost
    (D) a section of one gene repeats several times
    (E) a noncoding sequence of nucleotides is spliced from a chromosome

34. When the white peppered moth is better camouflaged against the white bark of the birch tree, the white moths are selected less often as food for the black moths. Over time the white moth frequency would increase due to

    (A) sexual selection
    (B) stabilizing selection
    (C) directional selection
    (D) diversifying selection
    (E) neutral selection

35. Polygenic inheritance is most evident when

    (A) black hair and brown hair produces brindle (black and brown hairs) in doges
    (B) hemophilia occurs more often in males than females
    (C) eye color in humans presents a great range of phenotypes between bright blue and deep brown
    (D) calico coloring occurs only in female cats
    (E) pink snapdragons are produced from white and red parents

36. All of the following organelles are paired with their correct cellular function EXCEPT

    (A) ribosomes—protein synthesis
    (B) mitochondrion—cellular respiration
    (C) chloroplast—photosynthesis
    (D) lysosome—storage
    (E) cytoskeleton—shape and support

37. Speciation occurs in animals when there is some level of reproductive isolation. This isolation can occur in many forms. Which of the following examples describes some level of mechanical isolation?

    (A) Two cicadia population mate during different months
    (B) Two dog breeds have anatomical incompatibilities to breeding
    (C) Some snails in a population mate on the rock surfaces and others under water
    (D) The offspring of a successful mating are weak
    (E) Some apple maggots mate on apple trees and some mate on Hawthorne trees

38. In order for a population to maintain a Hardy-Weinberg equilibrium, all of the following must be true EXCEPT

    (A) no natural selection can occur
    (B) the population size must be large
    (C) mating is completely random
    (D) individuals move freely in and out of the population
    (E) no net mutations must arise

39. Which is a correct pairing of nucleotides between the two strands of the DNA double helix?

    (A) cytosine with thymine
    (B) adenine with guanine
    (C) adenine with cytosine
    (D) guanine with thymine
    (E) cytosine with guanine

40. Alpha helices and beta pleated sheets are reoccurring patterns in the

    (A) DNA double helix
    (B) Primary structure of a protein
    (C) Secondary structure of a protein
    (D) Phosphate sugar bonds in DNA
    (E) Ring structure of a steroid

41. All of the following are lipids EXCEPT

    (A) monosaccharides
    (B) steroids
    (C) fats
    (D) phospholipids
    (E) waxes

42. The fluid mosaic model of the cell membrane includes all of the following facets EXCEPT

    (A) phospholipid bilayer
    (B) pntegral proteins that cross the membrane layers
    (C) cholesterol molecules
    (D) oligosaccharide surface markers
    (E) RNA side chains

43. All of the following are correct pairings of animals with their major respiratory organ EXCEPT

    (A) jellyfish; gills
    (B) clam; gills
    (C) cricket; trachiole
    (D) cow; pulmonary lung
    (E) spider; book lung

44. Birds have the same number of heart chambers as

    (A) fish
    (B) amphibians
    (C) reptiles
    (D) mammals
    (E) earthworms

45. Amniote eggs are found in all of the following EXCEPT

    (A) alligator
    (B) shark
    (C) penguin
    (D) kangaroo
    (E) pstrich

46. An interspecific interaction where one species benefits while the other is negatively affected is seen between

    (A) two squirrel species competing for the same nuts
    (B) a wasp species laying its eggs in the body of caterpillars
    (C) bees pollinating flowers while getting nectar
    (D) barnacles attaching to clam shells for movement
    (E) termites and the wood-digesting protist in their gut

47. Chloroflourocarbons are dangerous to the environment because they

    (A) damage the retinas of birds
    (B) collect in sediments at the bottom of streams
    (C) biologically magnify in the guts of large fish
    (D) breakdown ozone molecules in the atmosphere
    (E) absorb extra heat from the sun

48. During the formation of the primary RNA transcript from the DNA template, new nucleotides can only be added to one point on the RNA strand. That point is

    (A) at a specialized RNA binding site
    (B) at the gene operator
    (C) on the 3' end of the new RNA strand
    (D) between a sequence of repetitive nucleotides
    (E) at the telomeres

49. All of the following cellular functions are correctly matched to the the organelle responsible EXCEPT

    (A) ribosome; translation
    (B) mitochondrion; photosynthesis
    (C) lysosome; hydrolytic enzymes break up foreign materials
    (D) smooth ER; reduce toxicity
    (E) golgi complex; package materials to be sent from cell

Directions: There are five lettered headings for each group of questions, followed by a list of phrases or sentences. For each phrase or sentence, select the best answer. Answers may be used once, more than once or not at all in each group.

Use the following enzymes to answer questions 50-55.

    (A)   helicase
    (B)   DNA polymerase
    (C)   RNA polymerase
    (D)   DNA ligase
    (E)   reverse transcriptase

50.   Unwinds the DNA during transcription

51.   Unwinds the DNA during DNA replication

52.   Connects Okazaki fragments together

53.   Builds the RNA transcript during transcription

54.   Makes DNA from mRNA transcript

55.   Builds the DNA strands during replication

For questions 56-60, select the part of the alimentary canal that applies from the provided choices.

    (A)   pharynx
    (B)   esophagus
    (C)   duodenum
    (D)   jejunum
    (E)   Large intestine

56. Joint passage for air and food

57. Site of most enzymatic hydrolysis during digestion

58. Area with many villi and microvilli

59. Absorption of water site

60. Emulsification of lipids site

For questions 61-65, select one of the following modes of inheritance.

   (A) Polygenic inheritance
   (B) Incomplete dominance
   (C) Co-dominance
   (D) Sex-linkage
   (E) Linked genes

61. Wide range of phenotypes with no distinct categories

62. Heterozygotes have a trait that is intermediate to the two homozygote traits

63. Heterozygotes express both alleles

64. A and B blood type alleles

65. Inheritance patterns may be different in males and females

For questions 66-70, select one of the following animal phyla.

   (A) Annelida
   (B) Arthropoda
   (C) Echinodermata
   (D) Cnidaria
   (E) Chordata

66. Postanal tail and pharyngeal gill slits

67. Jointed appendages and exoskeletons

68. Radial adults, but bilateral juveniles

69. Radial adults and juveniles

70. Invertebrate deuterostome

For questions 71-75, select the most appropriate floral part from the ones provided.

    (A)   anther
    (B)   filament
    (C)   stigma
    (D)   style
    (E)   sepal

71. Site where pollen grains are produced

72. Site where pollen grains attach during fertilization

73. Pollen tube extends its length to reach the ovule

74. Protects the developing floral parts while in bud state

75. Site of male gametophyte tissue

Analyze the following graph that shows the reaction rate of an enzyme-catalyzed reaction as the temperature is changed. Answer the questions that follow.

The Effects of Temperature on an Enzyme-Catalyzed Reaction

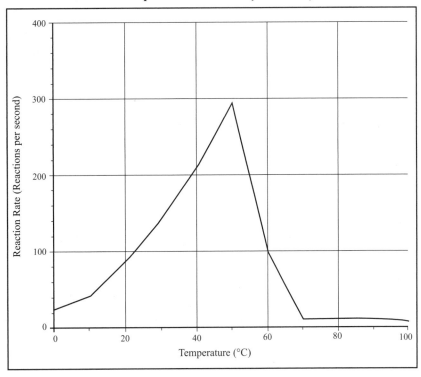

76. The temperature at which the reaction proceeds most quickly is

    (A) 25 °C
    (B) 40 °C
    (C) 50 °C
    (D) 70 °C
    (E) 100 °C

77. Which situation most appropriately describes the enzyme and substrate between 20 and 40 °C?

    (A) A competitive inhibitor is blocking the enzyme's active site and preventing enzyme-substrate binding
    (B) The pH has become too acidic for efficient enzyme functioning due to enzyme denatruation
    (C) The enzyme and substrate molecules are moving faster increasing the rate of collision and binding
    (D) A noncompetitive inhibitor is binding to the enzyme outside of the active site
    (E) The enzyme concentration is increasing

78. What is the most likely cause of the curve shape at 70 °C?

    (A) The enzyme has become denatured
    (B) The enzyme and substrate are moving more quickly
    (C) Inhibitors are blocking the enzyme action
    (D) The enzyme concentration has increased
    (E) The substrate concentration has increased

79. Enzymes are generally

    (A) carbohydrates
    (B) nucleic acids
    (C) lipids
    (D) steroids
    (E) proteins

Analyze the graph that shows the respiration rates of two groups of mice and crickets when exposed to different temperatures to answer the questions that follow.

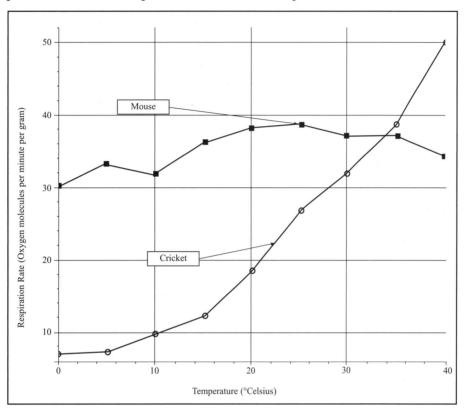

80. Which statement is true about an increase in temperature?

 (A) An increase in temperature results in a direct increase in mouse respiration rates
 (B) An increase in temperatures results in a direct increase in cricket respiration rates
 (C) Cricket respiration rates are indirectly proportional to temperature
 (D) Mouse respiration rates are indirectly proportional to temperature
 (E) Temperature has the same effect on the respiration rates of cricket and mice

81. If this graph was extrapolated to 95 degrees Celsius,

 (A) the respiration rate of crickets will continue to increase at the same rate
 (B) the respiration rate of mice will increase at a rate similar to the crickets
 (C) the respiration rates of both species would slow dramatically
 (D) the respiration rate of the crickets would be unaffected
 (E) the respiration rate of the mice would continue to remain constant

82. The cricket is affected by temperature to a greater degree than the mouse because the cricket

    (A) is ectothermic
    (B) uses trachioles to respire
    (C) has a smaller body size
    (D) has an exoskeleton
    (E) lives naturally in a warmer environment

An engineered plasmid was inserted into bacteria through transformation. The bacteria that were exposed to the engineered plasmid are labeled as + bacteria and the bacteria that were not exposed to the plasmid were labeled as – bacteria. Both types of bacteria were grown on agar plates containing only luria broth or plates containing luria broth and ampicillin. The growth of the bacteria is presented in the following table.

Luria Broth Agar	**Luria Broth and Ampicillin in Agar**

I
 	
    + plasmid: Bacterial Growth
	III
	
	    + plasmid: Bacterial growth

II
 	
    – plasmid: Bacterial Growth
	IV
	    – plasmid: No growth of bacteria

83. The most probable explanation for the resultant bacterial growth patterns is that

    (A) the engineered plasmid was not successfully inserted into any of the bacteria
    (B) the engineered plasmid contained a gene for ampicillin resistance
    (C) the engineered plasmid caused the transformed bacteria to fluoresce
    (D) conjugation occurred between the + bacteria and the – bacteria
    (E) an operon controls the gene for ampicillin resistance

84. Why did bacteria grow successfully on plate II (the luria broth plate with the – plasmid bacteria)?

    (A) There was nothing in the environment to hinder bacterial growth
    (B) These bacteria were successfully transformed in the experiment
    (C) These bacteria are resistant to ampicillin
    (D) These bacteria are expressing a human gene that was inserted
    (E) It is the only plate where the bacteria are capable of successful reproduction

85. If the engineered plasmid also contained a gene for making human growth hormone, which plate would you use to be sure that you have bacteria capable of making human growth hormone?

    (A) I only
    (B) I and II
    (C) I and III
    (D) III only
    (E) IV only

86. Engineered plasmids are one way to artificially alter the genome of a bacterium. Bacteria can naturally alter their genomes by all of the following EXCEPT

    (A) transformation
    (B) transduction
    (C) conjugation
    (D) mutation
    (E) histone acetylation

Unknown Animal Number	Traits of Discovered Animal
1	Exoskeleton, jointed appendages, 8 pairs of appendanges, head and thorax region
2	Exoskeleton, exhibits torsion, spiral shell, hermaphroditic
3	Fur, large eyes, mammary glands, developing young retained in body of the mother
4	Keritin scales, leathery shelled eggs, endoskeleton
5	Beak, feathers, calcium carbonate eggs, well-defined vision

A man that is heterozygous for A blood type and is heterozygous for the Rh antigen (Rh antigen is inherited through simple dominance and recessiveness where positive is dominant and negative is recessive) marries a woman that is homozygous for B blood type and is Rh negative.

87. What percentage of the offspring is likely to have O blood type?

    (A) 0%
    (B) 25%
    (C) 50%
    (D) 75%
    (E) 100%

88. What percentage of the offspring is likely to be Rh negative?

    (A) 0%
    (B) 25%
    (C) 50%
    (D) 75%
    (E) 100%

89. What percentage of the offspring is likely to have AB positive blood?

    (A) 0%
    (B) 25%
    (C) 50%
    (D) 75%
    (E) 100%

90. What percentage of the offspring will have the same phenotype and genotype as the father?

    (A) 0%
    (B) 25%
    (C) 50%
    (D) 75%
    (E) 100%

Five new species of terrestrial animals were discovered in a hidden cave. Some traits for the animals are presented in the following table.

91. How many of the newly discovered animals probably are invertebrates?

    (A) 1
    (B) 2
    (C) 3
    (D) 4
    (E) 5

92. What is another trait that is likely to be true of species 1?

    (A) pulmonary lung
    (B) gills
    (C) trachioles
    (D) book lung
    (E) gastrovascular cavity

93. Species 4 belongs to the Phylum

    (A) Cnidaria
    (B) Arthropoda
    (C) Echinodermata
    (D) Chordata
    (E) Annelida

94. A radula for feeding is probably found on animal

    (A) 1
    (B) 2
    (C) 3
    (D) 4
    (E) 5

95. A mouse is most closely related to species

    (A) 1
    (B) 2
    (C) 3
    (D) 4
    (E) 5

96. Which one of the following organisms would be most closely related to species 2?

    (A) clam
    (B) earthworm
    (C) snake
    (D) alligator
    (E) starfish

The transpiration rates of plant cuttings were monitored as the plant cuttings were exposed to various environmental features.

Environment of Plant Cutting	Average Transpiration Rate (ml/minute)
Presence of wind	3.89
Extra light	3.25
Humidity	0.08
Darkness	0.12
Normal growing conditions	1.20

97. Transpiration is best described as the

    (A) pressure exerted on water in the plant as water and minerals move into the roots
    (B) evaporation of water through the leaf tissue
    (C) tendency of water molecules to be attracted to one another
    (D) tendency of plant cells to avoid freezing as temperatures decline
    (E) rate of sugar production in a plant

98. The plant cells that most directly affect the rate of transpiration are the

    (A) sclerenchyma cells
    (B) parenchyma cells
    (C) sieve tube members
    (D) collenchyma cells
    (E) guard cells

99. If there was another plant group that was exposed to both the wind and the extra light, transpiration rates for that group would be

    (A) between 1.20 and 3.25 ml/minute
    (B) between 0.12 and 1.20 ml/minute
    (C) greater than 3.89 ml/minute
    (D) between 3.25 and 3.89 ml/minute
    (E) below 3.25 ml/minute

100. The water being lost in these plants has moved through the plant through the

    (A) phloem
    (B) xylem
    (C) cork cambium
    (D) casparian strip
    (E) sieve tube members

FREE-RESPONSE QUESTIONS FOR AP PRACTICE TEST #2

Directions: Answer all questions

You must answer the questions in essay form (no outlines). You may use diagrams and drawings, but they must be accompanied by discussion. Read each question thoroughly before you begin to write.

1. In anatomy and physiology of organisms, it is said that form follows function. **Select** 3 of the following mammalian structures and **describe** how the form of that structure reflects its function.

   a. Nephron
   b. Neuron
   c. Sarcomere
   d. Stomach
   e. Kidney
   f. Heart

2. A DNA sample was collected at a crime scene to be analyzed forensically. Several processes were employed to forensically analyze the DNA sample. **Explain** any 3 of the following steps to processing that DNA sample.

   a. PCR
   b. Restriction digest
   c. Electrophoresis
   d. Analysis of VNTRs

3. The growth rate of populations tends to follow a predictable pattern. Many populations exhibit either exponential or logistic growth.

   a. **Define** and **explain** both exponential growth and logistic growth.
   b. **Describe** parameters of a population that tends to exhibit logistic growth.
   c. Draw a graph that shows logistic growth.

4. The current system used to classify organisms relies on three domains of life: Archae, Bacteria, and Eukarya.

   a. **List** and **describe** prominent features of each of the three domains.
   b. **List** and **describe** the prominent features of the major kingdoms.
   c. How has the advancement of molecular evidence caused discussion about the traditional taxonomic placement of some organisms?

DATA CORRELATION: Sample Exam #2

After you have completed your practice exam, circle the numbers of the items that you
missed. Use this information to guide you in determining which topics may require a
more in-depth review.

I.   Molecules and Cells

Photosynthesis	2,3
Cellular Respiration	23
Organelle Structure and Function	36,42,49
Molecules and Bonding	40,41
Enzymes	76,77,78,79

II.   Genetics and Evolution

Inheritance Patterns	4,5,35,61,62,63,64,65,87,88,89,90
Evolution	9,34,37,38
Molecular Genetics (DNA and RNA)	10,11,32,33,39,48,50,51,52,53,54,55
Nuclear Division	16,30
Biotechnology	18,19,83,84,85,86

III.   Organisms and Populations

Plant Structure and Function	1,15,22,24,25,71,72,73,74,75,97,98,99,100
Animal Structure and Function	12,13,14,20,21,26,27,43,44,45,56,57,58,59,60,80,81,82
Classification	28,29,31,66,67,68,69,70,91,92,93,94,95,96
Ecology	6,7,8,17,46,47

Sample Examination III

Directions: Each of the following questions or statements is followed by five possible answers or sentence completions. Select the best answer or completion for each question or statement.

1. All of the following statements regarding cancer cells are true EXCEPT

   (A) cancer cells do not exhibit anchorage inhibition.
   (B) cancer cells secrete signal molecules that cause blood vessels to grow towards a tumor.
   (C) cancer cells exhibit density dependent inhibition.
   (D) cancer cells stop at random points in the cell cycle instead of at checkpoints.
   (E) cancer can be caused by a virus.

2. The excretory mechanism found in arthropods are

   (A) contractile vacuoles
   (B) flame cells
   (C) nephridia
   (D) malphighian tubules
   (E) nephrons

3. Which of the following is an example of a negative feedback system?

   (A) control of uterine contractions and the release of oxytocin
   (B) control of calcium levels by parathyroid hormone and calcitonin
   (C) control of digestion by the conversion of pepsinogen to pepsin
   (D) control of blood clotting mechanisms
   (E) control of action potentials in the nervous system

4. Which of the following organisms has a closed circulatory system, bilateral symmetry, segmentation, and cephalization?

   (A) hydra
   (B) earthworm
   (C) sea anemone
   (D) roundworm
   (E) sponge

181

5. Which of the following types of bonds joins monomers in carbohydrates?

(A) peptide bond
(B) glycosidic linkage
(C) hydrogen bond
(D) ionic bond
(E) hydrophobic interaction

6. The first lines of defense in the mammalian immune system involve all of the following EXCEPT

(A) cilia
(B) lysozyme
(C) gastric juice
(D) antimicrobial proteins
(E) immunoglobulin

7. Which of the following would result from placing a plant cell in a hypertonic environment?

(A) the plant cell would become turgid from an influx of water.
(B) plasmolysis would occur.
(C) there would be no net movement of water into or out of the cell.
(D) the plant cell would have a lower water potential than that of the surrounding environment.
(E) the plant cell wall would burst due to the pressure of the surrounding environment.

8. Which of the following occurs during anaphase of mitosis?

(A) Separation of homologous chromosomes
(B) Disintegration of the nuclear membrane
(C) Formation of spindle fibers
(D) Separation of sister chromatids
(E) Crossing over of segments of homologous chromosomes

9. Which of the following plant hormones is responsible for slowing plant growth during periods of unfavorable environmental conditions?

(A) ethylene
(B) abscisic acid
(C) gibberellin
(D) cytokinin
(E) indolacetic acid (IAA)

10. Which of the following hormones causes a release of bile from the gallbladder?

    (A) gastrin
    (B) pepsin
    (C) cholesystekinin
    (D) insulin
    (E) glucagon

11. Which of the following adaptations can be seen in a desert mammal?

    (A) short loops of Henle
    (B) ammonia as an excretory waste
    (C) increased filtration at Bowman's capsule
    (D) secretion of waste as uric acid
    (E) urine that is hypotonic to body tissues

12. Which of the following is NOT an adaptation of plants in order to increase photosynthetic efficiency?

    (A) Opening of stomata at night instead of during the day
    (B) Incorporation of CO_2 into four carbon sugars
    (C) Export of CO_2 from the mesophyll to the bundle sheath cells
    (D) Elimination of photorespiration mechanisms
    (E) Adding of O_2 to the Calvin Cycle as CO_2 becomes scarce

13. Which of the following statements regarding photoperiodism is NOT correct?

    (A) Short day plants will flower if the duration of night is longer than the critical length.
    (B) Long day plants will not flower if the night is shorter than a critical dark period.
    (C) Long day plants will flower if the period of darkness is shortened by light.
    (D) If the nighttime part of the photoperiod is interrupted by light, the plants will not flower.
    (E) If the daytime portion of the photoperiod is broken by a brief exposure to darkness there is no effect on flowering.

14. Which of the following types of bond determines the secondary structure of a polypeptide?

    (A) hydrogen
    (B) covalent
    (C) ionic
    (D) van der Waals interactions
    (E) hydrophobic interactions

15. Which of the following functional groups is present in the backbone of all amino acids?

    (A) carbonyl
    (B) carboxyl
    (C) hydroxyl
    (D) methyl
    (E) phosphate

16. All of the following statements describe structural or functional differences between complex carbohydrates EXCEPT

    (A) a starch molecule is helical and a cellulose molecule is straight.
    (B) the glucose molecules in starch are in the alpha configuration, whereas the glucose molecules in cellulose are in the beta configuration.
    (C) starch is the storage form of carbohydrate in mammals.
    (D) cellulose is the structural form of carbohydrate in plants.
    (E) enzymes that digest starch are unable to digest cellulose because of structural differences.

17. The cell membrane is primarily composed of which of the following molecules

    (A) phospholipids
    (B) cholesterol
    (C) phosphates
    (D) steroids
    (E) glycerol

18. Which of the following molecules is a monomer?

    (A) glycogen
    (B) glucose
    (C) albumin
    (D) chitin
    (E) starch

19. All of the following substances could be found in cell walls EXCEPT

    (A) chitin
    (B) peptidoglycan
    (C) cellulose
    (D) glycogen
    (E) proteins

20. Which of the following organelles are present in a prokaryotic cell?

    (A) mitochondrion
    (B) endoplasmic reticulum
    (C) nucleolus
    (D) chloroplast
    (E) ribosome

21. All of the following are differences between aerobic respiration and fermentation EXCEPT

    (A) both pathways use glycolysis to oxidize glucose.
    (B) the final electron acceptor in respiration is pyruvate.
    (C) NAD^+ is the oxidizing molecule for respiration and fermentation.
    (D) fermentation occurs in the absence of oxygen, whereas aerobic respiration occurs in the presence of oxygen.
    (E) respiration can yield much more ATP than fermentation.

22. All of the following statements are true regarding photosynthesis EXCEPT

    (A) the production of oxygen occurs during the light dependent reactions in the chloroplast.
    (B) the Calvin cycle can run in the dark.
    (C) the electron transport chain pumps protons across a membrane from a region of high concentration to low concentration and then diffuse back across, driving ATP synthesis.
    (D) the light reactions capture energy in the form of NADPH and ATP to be used in the Calvin cycle.
    (E) the Calvin cycle incorporates one carbon dioxide molecule with a five-carbon sugar, in a reaction catalyzed by the enzyme rubisco or RuBP.

23. Water moderates air temperature due to the fact that

    (A) water absorbs or releases a large amount of heat and increases considerably in temperature.
    (B) water has an unusually low specific heat.
    (C) when the temperature of water drops, many additional hydrogen bonds form, releasing energy in the form of heat.
    (D) it has a very low heat of vaporization that keeps temperatures very steady.
    (E) the covalent bonds between hydrogen and oxygen are very strong and do not break very easily.

24. All of the following are true about the transcription and processing of an eukaryotic gene EXCEPT

    (A) the DNA contains a promoter region where transcription begins.
    (B) removal of noncoding DNA sequences called exons occurs prior to translation.
    (C) a 5' cap and poly-A tail is added to the primary transcript.
    (D) transcription is initiated by RNA polymerase and transcription factors.
    (E) some transcription factors function as repressors to inhibit expression of a specific gene.

25. What is the probability that parents who are heterozygous for three genes will produce offspring that are also heterozygous, assuming independent assortment of all gene pairs?

    (A) 1/4
    (B) 1/8
    (C) 1/2
    (D) 1/16
    (E) 1/6

26. Sickle cell anemia is an inherited disease caused by a recessive allele. If a woman with sickle cell anemia and a man who is a carrier have two children, what are the chances that both children will have sickle cell anemia?

    (A) 1/2
    (B) 1/4
    (C) 1/6
    (D) 1/3
    (E) 1/8

27. Which of the following types of mutations is displayed in the following chromosome?

    A B C D E F G ⟶ A B E D C F G

    (A) deletion
    (B) duplication
    (C) inversion
    (D) translocation
    (E) nondisjunction

28. Which of the following statements is NOT true regarding the following equation?

$$C_6H_{12}O_6 + O_2 \rightarrow 6\ CO_2 + 6\ H_2O$$

   (A) This reaction is exergonic.
   (B) The products store less energy than the reactants.
   (C) This reaction occurs spontaneously.
   (D) $C_6H_{12}O_6$ becomes oxidized in this reaction.
   (E) CO_2 is reduced in this reaction.

29. Cellular respiration is NOT controlled by

   (A) allosteric enzymes that regulate glycolysis
   (B) the inhibition of phosphofructokinase by ATP
   (C) ATP concentrations within the cell.
   (D) the stimulation of enzymes due to an increase in AMP levels.
   (E) the production of citrate during glycolysis.

30. The light reaction of photosynthesis supply the Calvin cycle with

   (A) ATP and NADPH
   (B) NADP and H_2O
   (C) CO_2 and NADPH
   (D) $C_6H_{12}O_6$ and NADP
   (E) O_2 and CO_2

31. Which of the following enzymes is found in retroviruses that produce DNA from an RNA template?

   (A) DNA ligase
   (B) DNA polymerase
   (C) RNA polymerase
   (D) reverse transcriptase
   (E) primase

32. Which of the following is NOT true of the Calvin cycle

   (A) The process occurs in the stroma
   (B) The process uses ATP
   (C) $NADP^+$ is a product
   (D) CO_2 is converted to glucose
   (E) Oxygen is released

33. In a DNA molecule, hydrogen bonds would most likely be found between

    (A) a phosphate molecule and a sugar molecule
    (B) two sugar molecules
    (C) a sugar molecule and a nitrogen base
    (D) two nitrogenous bases
    (E) two phosphate molecules

34. Which of the following enzymes synthesizes the leading strand during DNA replication?

    (A) helicase
    (B) RNA polymerase
    (C) topoisomerase
    (D) DNA ligase
    (E) DNA polymerase

35. Which of the following terms identifies a change in genotype due to the uptake of foreign DNA by a cell?

    (A) translocation
    (B) transformation
    (C) translation
    (D) transduction
    (E) transcription

36. Which of the following pairs does NOT illustrate the concept of homologous structures?

    (A) The arm of a human and the forelimb of a cat
    (B) The wing of a bat and the forelimb of a whale
    (C) The wing of a bird and the wing of an insect
    (D) The gills of a fish and the throat of a human
    (E) The tail of a chimpanzee and the tailbone of a human

37. Which of the following conditions is NOT true of the Hardy-Weinberg equilibrium?

    (A) The smaller the population the greater the fluctuation of gene frequency.
    (B) Mutations are necessary to maintain genetic equilibrium.
    (C) The immigration of individuals to a population can affect gene frequency.
    (D) Differential reproductive success can impact allele frequency.
    (E) The introduction of genes into a population modifies the gene pool.

38. Over many generations, the color of field mice gradually darkens in response to environmental changes in rock color. This is an example of

    (A) directional selection
    (B) founder effect
    (C) disruptive selection
    (D) stabilizing selection
    (E) balanced polymorphism

39. Which of the following terms best describes the concept that even in closely related species of plants, the flowers have very different appearances that attract different pollinators?

    (A) habitat isolation
    (B) temporal isolation
    (C) mechanical isolation
    (D) hybrid isolation
    (E) gametic isolation

40. Which of the following is NOT a prezygotic barrier?

    (A) Two species of snakes live in the same area, but one is aquatic and the other is terrestrial.
    (B) Two species of squirrels live in the same area, but one species mate in the winter while the other species mate in the summer.
    (C) Two species of birds have different mating songs.
    (D) Sperm and egg from different species of sea urchins are unable to fuse in the water.
    (E) The hybrid offspring of a donkey and a horse is sterile.

41. Which of the following terms includes the others?

    (A) domain
    (B) family
    (C) kingdom
    (D) order
    (E) species

42. Which of the following is NOT present in organisms that belong to Domain Bacteria?

    (A) peptidoglycan cell walls
    (B) circular chromosomes
    (C) introns
    (D) RNA polymerase
    (E) histones associated with DNA

43. Which of the following organisms would NOT be grouped with other protists?

    (A) diatoms
    (B) slime molds
    (C) water mold
    (D) protozoans
    (E) moss

44. Which of the following is characteristic of monocots?

    (A) netlike veins
    (B) vascular tissue arranged in a ring
    (C) fibrous root system
    (D) two cotyledons
    (E) flowers in multiples of four or five

45. An insect lays eggs on a living host and then the larvae feed on the body of the host, eventually killing it. This is an example of

    (A) evolution
    (B) parasitism
    (C) commensalism
    (D) mutualism
    (E) predation

46. All of the following statements regarding speciation are true EXCEPT

    (A) species can diverge in spurts of relatively rapid change.
    (B) species may undergo most of their modifications as they first diverge from their parent species.
    (C) species may descend from a common ancestor and gradually diverge more and more as they acquire different adaptations.
    (D) through diversifying selection, a species is adapted to an environment that stays the same, and stasis may result.
    (E) speciation can follow the emergence of a geographic barrier.

47. All of the following statements regarding reproductive immunology are true EXCEPT

    (A) the protective layer called the trophoblast prevents the embryo from contacting maternal tissue.
    (B) the trophoblast produces a chemical signal that induces the production of a white blood cell in the uterus that acts as a suppressor and prevents other white blood cells from attacking the embryo.
    (C) the trophoblast does not develop from the blastocyst and therefore does not have foreign markers.
    (D) similarities in cellular markers of the mother to the father may result in a decreased immune response and possible spontaneous abortion of the embryo.
    (E) a weak immune response may account for multiple miscarriages in women.

48. Which of the following hypotheses best explains how epiphytes obtain water?

    (A) They absorb water from the air.
    (B) They absorb water through the tissues of the trunk.
    (C) Their roots grow downward from the canopy and absorb water from the soil.
    (D) They have stomata on their roots.
    (E) They absorb water from the mesophyll of the leaves of the tree.

49. Which of the following are horizontal stems that enable a strawberry plant to reproduce asexually?

    (A) stolons
    (B) bulbs
    (C) tubers
    (D) rhizomes
    (E) axillary buds

50. Which of the following statements is true of mature phloem tissue?

    (A) It contains nuclei and ribosomes.
    (B) It is composed of tracheids and vessel members.
    (C) It is dead upon maturity.
    (D) It provides mechanical support to the plant.
    (E) It functions in the conduction of sugar.

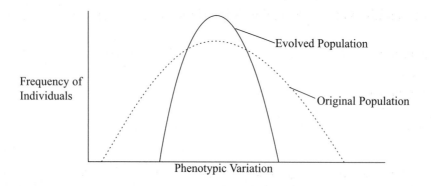

51. Which of the following patterns could be best explained by the above diagram?

    (A) divergent evolution
    (B) diversifying selection
    (C) directional selection
    (D) stabilizing selection
    (E) analogous evolution

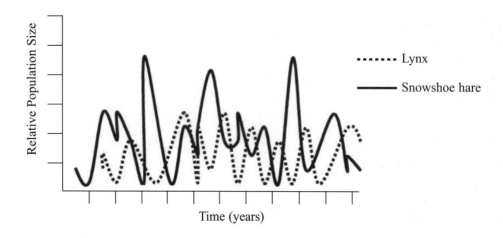

52. All of the following statements are true of the above graph EXCEPT

    (A) the fluctuation in hare population size could be due to the available food supply.
    (B) a spike in the hare population corresponds with a decrease in lynx population.
    (C) predation by lynx may limit hare populations.
    (D) rainfall levels may affect the food supply of the hare impact population size.
    (E) the snowshoe hare will eventually become extinct due to predation by the lynx.

53. Which of the following terms includes all of the others?

   (A) population
   (B) species
   (C) community
   (D) ecosystem
   (E) habitat

54. Which of the following will decrease the amount of dissolved oxygen in water?

   (A) A decrease in temperature.
   (B) An increase in salt concentration.
   (C) An increase in photosynthetic activity.
   (D) A decrease in respiratory rate.
   (E) An increase in circulation.

55. Organisms from which of the following phyla do NOT have a coelom?

   (A) Arthropoda
   (B) Mollusca
   (C) Annelida
   (D) Platyhelminthes
   (E) Vertebrata

56. Which of the following phyla contains organisms that have radial cleavage and the mouth forms from a secondary opening?

   (A) Mollusca
   (B) Annelida
   (C) Arthropoda
   (D) Echinodermata
   (E) Rotifera

57. The movement of water from the roots to the leaves of a tracheophyte can best be attributed to

   (A) capillary action
   (B) gravitational pull
   (C) turgor pressure
   (D) root pressure
   (E) transpiration

58. Which of the following pH levels would promote the highest rate of reaction for trypsin, an intestinal enzyme?

    (A) pH 2
    (B) pH 4
    (C) pH 5
    (D) pH 8
    (E) pH 12

59. All of the following are examples of countercurrent exchange EXCEPT

    (A) oxygen exchange in the gills of fish
    (B) heat transfer in the legs of a Canadian goose
    (C) solute concentration in the mammalian kidney
    (D) temperature regulation in the swimming muscles of a shark
    (E) nutrient absorption in the small intestine

60. All of the following statements are true EXCEPT

    (A) endotherms need to consume more food than ectotherms of the same size.
    (B) ectotherms have lower body temperatures than endotherms.
    (C) endotherms can perform activities for a much longer period of time than endotherms.
    (D) ectotherms tolerate greater fluctuations in temperature than endotherms.
    (E) endotherms have higher metabolic rates than ectotherms.

61. Which of the following events does NOT occur during a skeletal muscle contraction?

    (A) A crossbridge is formed between actin and myosin.
    (B) Ca^{2+} is released from the sarcoplasmic reticulum
    (C) A motor neuron causes the release of a neurotransmitter
    (D) The troponin-tropomyosin complex is repositioned
    (E) Ca^{2+} transmits actions from the neuron to the muscle fiber

Directions: There are five lettered headings for each group of questions, followed by a list of phrases or sentences. For each phrase or sentence, select the best answer. Answers may be used once, more than once or not at all in each group.

Questions 62-66 refer to the following diagram.

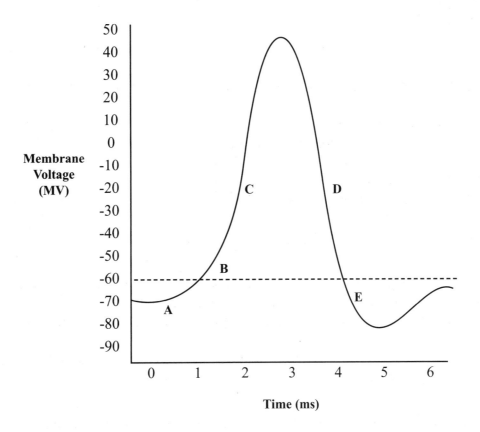

62. The point at which a neuron is unable to fire.

63. The threshold value where a neuron begins an "all or nothing' response.

64. An influx of Na+ causes depolarization.

65. An efflux of K+ causes repolarization.

66. Hyperpolarization of the neuron occurs.

Questions 67-69

    (A) mitochondria
    (B) nucleolus
    (C) endoplasmic reticulum
    (D) ribosome
    (E) Golgi apparatus

67. The site of protein synthesis.

68. The site of aerobic respiration.

69. Enzymes are produced here.

Questions 70-73

    (A) fern
    (B) moss
    (C) conifer
    (D) algae
    (E) flowering plant

70. Contain seeds not enclosed in chambers.

71. A nonvascular plant

72. Can be monocot or dicot.

73. Cone bearing plant

Questions 74-77

    (A) taiga
    (B) desert
    (C) deciduous forest
    (D) tropical rainforest
    (E) grassland

74. Characterized by coniferous forests and very cold winters.

75. Epiphytes commonly grow on the trees.

76. Deeply rooted shrubs can be found here.

77. Vegetation of predominantly tall trees with little growth on the forest floor.

Questions 78-81

    (A) mRNA
    (B) tRNA
    (C) rRNA
    (D) zDNA
    (E) cDNA

78. Synthesized from mRNA

79. Able to recognize mRNA codons

80. Moves from the nucleus to the ribosome during protein synthesis

81. Specifies the primary structure of a protein

Questions 82-83 refer to the following food chain.

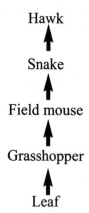

Hawk
↑
Snake
↑
Field mouse
↑
Grasshopper
↑
Leaf

82. Which of the following terms best describes the trophic level of the field mouse?

    (A) primary producer
    (B) primary consumer
    (C) secondary consumer
    (D) tertiary consumer
    (E) quaternary consumer

83. Which of the following organisms from the food web has the lowest biomass in the ecosystem?

    (A) field mouse
    (B) grasshopper
    (C) hawk
    (D) snake
    (E) leaf

Questions 84-85 refer to the following diagram of DNA fragments from five samples separated by gel electrophoresis.

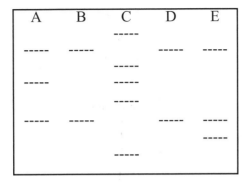

84. Which of the following statements is NOT true regarding the movement of the DNA fragments through the gel?

 (A) The smaller fragments will travel a greater distance.
 (B) The fragments move towards the negative end of the gel.
 (C) Larger fragments travel slower throughout the gel.
 (D) An electrical potential is created to cause movement of the fragments in the gel.
 (E) Fragments of DNA will separate based upon their size.

85. Which of the following statements does NOT describe the results of the gel?

 (A) Samples B and D are similar DNA sequences.
 (B) Sample C produced the most DNA fragments.
 (C) Samples C and D have little in common
 (D) Sample C produced the smallest DNA fragment.
 (E) Sample A contains the largest DNA fragment.

Questions 86-88

   I.   Mitosis
  II.   Meiosis I
 III.   Meiosis II

86. Sister chromatids separate and move to the poles.

 (A) I only
 (B) II only
 (C) I and III only
 (D) II and III only
 (E) I, II and III

87. Homologous chromosomes separate.

     (A) II only
     (B) III only
     (C) II and III only
     (D) I and III only
     (E) I, II and III

88. Tetrads are formed.

     (A) I only
     (B) II only
     (C) III only
     (D) II and III only
     (E) I and III only

Questions 89-90 refer to the following diagram.

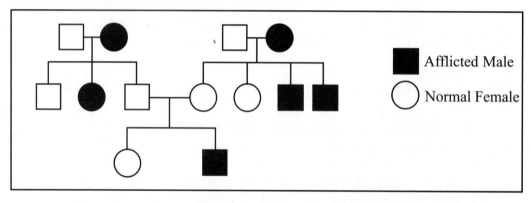

89. This pedigree is best explained by the following type of inheritance pattern?

     (A) X linked recessive
     (B) Autosomal dominant
     (C) Autosomal recessive
     (D) X linked dominant
     (E) Y linked recessive

90. Based on the type of inheritance, what disorder could be displayed by the pedigree

     (A) cystic fibrosis
     (B) achondroplasia
     (C) Huntington's Disease
     (D) polyploidy
     (E) color blindness

Questions 91-93 refer to an experiment in which a dialysis bag filled with a solution of 0.3 M sucrose was placed in a beaker containing 0.5 M sucrose as shown in the diagram below.

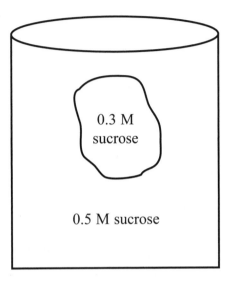

91. The solution in the beaker

   (A) is hypotonic to the dialysis bag
   (B) has a lower solute potential than the dialysis bag
   (C) is in equilibrium with the dialysis bag
   (D) has a lower water potential than the dialysis bag
   (E) is isotonic to the dialysis bag

92. After the dialysis bag remains in the beaker overnight,

   (A) the dialysis bag will become turgid
   (B) the concentration of sucrose outside the bag will be 0.4 M
   (C) sucrose will move from the beaker to the dialysis bag
   (D) the dialysis bag will lose water
   (E) the solution in the dialysis bag will become hypertonic to the solution in the beaker.

93. If, after 24 hours, the concentration of sucrose within the dialysis bag and the beaker is equal, which of the following is NOT a reasonable conclusion?

   (A) Sucrose is able to fit through the pores of the dialysis bag.
   (B) The dialysis bag contains a selectively permeable membrane.
   (C) There was a net movement of water out of the dialysis bag.
   (D) The water potential inside the bag was lower than that of the surrounding environment.
   (E) The water potential outside the bag is now equal to that of the dialysis bag.

Questions 94-95

   I.   facilitated diffusion
   II.   osmosis
   III. active transport

94. Which of the following do not require an expenditure of energy?

   (A) I and II
   (B) II and III
   (C) III only
   (D) I only
   (E) I, II and III

95. Which of the following transports solutes against a concentration gradient?

   (A) I and II
   (B) I and III
   (C) III only
   (D) I only
   (E) I, II and III

Questions 96-97 refer to an experiment that was conducted to measure the effect of light on photosynthetic rate. Three different treatments were used together with an indicator, DPIP, to determine the rate of photosynthesis. DPIP acts as an electron acceptor and is blue in the oxidized state and clear in the reduced state, after accepting electrons in photosynthesis.

    I.   Boiled and exposed to light
   II.   Not boiled and exposed to light
  III.   Not boiled and not exposed to light

96. Which of the following statements regarding this experiment can you reasonably expect to be true?

    (A)   Boiling of the chloroplasts in Environment I will cause the lowest transmittance.
    (B)   The chloroplasts in Environment III will have an increase in the amount of DPIP reduced.
    (C)   Electrons of the pigment molecules in the chloroplast of Environments I and II will be energized (reduced) and move via energy carriers to generate ATP.
    (D)   The photosynthetic rate in Environment II will be the lowest of the three.
    (E)   Increased temperature in Environment I will cause the greatest photosynthetic rate of the three environments.

97. Boiling of the chloroplasts

    (A)   disrupts the thylakoid membranes
    (B)   causes increased transmittance
    (C)   increases photosynthetic activity
    (D)   causes DPIP to be reduced
    (E)   increases enzymatic activity

Questions 98-100 refer to the following illustration of a segment of DNA and reaction sites for two restriction enzymes, A and B, as well as the lengths of the DNA fragments produced.

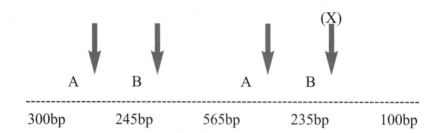

98. The resulting double stranded fragments have at least one single stranded end which is called a

    (A) sticky end
    (B) RNA fragment
    (C) plasmid
    (D) RFLP
    (E) vector

99. If the DNA were only treated with restriction enzyme A, which of the following would result?

    (A) Five segments of DNA with lengths of 300bp, 245bp, 565bp, 235bp and 100bp
    (B) Three segments of DNA with lengths of 300bp, 810bp, and 335bp
    (C) Two segments of DNA with lengths of 335bp and 1110bp
    (D) Three segments of DNA with lengths of 545bp, 800bp, and 100bp
    (E) Four segments of DNA with lengths of 300bp, 245bp, 565bp and 335bp

100. If a mutation occurred at site (X) of the DNA and the DNA was treated with both restriction enzymes A and B all of the following could occur EXCEPT?

    (A) Restriction enzyme B would not recognize the sequence of DNA at site (X) and would not cut.
    (B) Fragment size could change
    (C) Four DNA fragments could be produced
    (D) Enzyme B would become inactivated
    (E) Restriction enzyme A's action would be unaffected.

Sample Examination #3: Free Response Questions

Directions: Answer all questions.

You must answer the questions in essay form (no outlines). You may use diagrams and drawings, but they must be accompanied by discussion. Read each question thoroughly before you begin to write.

1. Osmosis is a driving force for homeostasis in plants and animals. **Define** osmosis and **discuss** how the movement of water regulates THREE of the following activities:

   a.   water and solute transport through plants
   b.   solute concentration in the mammalian kidney
   c.   stomatal opening in leaves
   d.   blood pressure in humans

2. Interdependence between species in nature is essential for maintaining function at all levels of organization.

   a.   **Discuss** the role of prokaryotes in chemical cycling in plant roots or the animal digestive tract.
   b.   **Explain** how interdependence maintains trophic levels in ecosystem.

3. Proteins are important for proper biological functioning.

   a.   **Discuss** the levels of organization of proteins.
   b.   **Describe** three possible functions of proteins in the human body.
   c.   **Explain** how a mutation can change the function of a protein.

4. The cell cycle functions in regulating information flow through organisms.

   a.   **Discuss** how alleles are distributed by the process of meiosis to the gametes.
   b.   **Explain** how the cell cycle is controlled.
   c.   **Explain** how these controls prevent errors in inheritance.

DATA CORRELATION: Sample Examination 3

After you have completed your practice exam, circle the numbers of the items you missed. Use this information to guide you in determining which topics may require a more in-depth review.

I.   Molecules and Cells

Bonding, Water, pH	7, 23
Organic Chemistry, Biological Molecules	5, 14, 15, 16, 18
Free Energy Changes, Enzymes	28, 96
Cell Structure	19, 20, 67, 68, 69
Cell Membranes	7*, 17, 91, 92, 93, 94, 95
Cell Cycle	1, 8
Respiration	21, 29, 32
Photosynthesis	22, 30, 97

II.   Genetics and Evolution

Meiosis	84, 85, 86, 87, 88
Inheritance patterns	25, 26, 89, 90
Molecular Genetics (RNA/DNA)	24, 27, 31, 33, 35, 78, 79, 80, 81
Biotechnology	36, 84, 85, 98, 99, 100
Evolution	37, 38, 39, 40, 46, 51

III.   Organisms and Populations

Phylogenetic classification	41, 42, 43, 55, 56
Plant SF	9, 12, 44, 48, 49, 50, 57, 70, 71, 72, 73
Animal SF	2, 3, 4, 6, 10, 11, 13, 47, 54, 58, 59, 60, 61, 62, 63, 64, 65, 66
Ecology/Behavior	45, 52, 53, 74, 75, 76, 77, 82, 83

NOTES

NOTES

NOTES

NOTES